P9-DUC-558

WITHDRAWN

MEASURING THE UNIVERSE

*

MEASURING THE UNIVERSE

Cosmic Dimensions from Aristarchus to Halley

*

ALBERT VAN HELDEN

THE UNIVERSITY OF CHICAGO PRESS * CHICAGO & LONDON

Albert Van Helden is professor of history at Rice University.

THE UNIVERSITY OF CHICAGO PRESS, CHICAGO 60637
THE UNIVERSITY OF CHICAGO PRESS, LTD., LONDON

Printed in the United States of America

94 93 92 91 90 89 88 87 86 85   54321

LIBRARY OF CONGRESS CATALOGING IN PUBLICATION DATA

Van Helden, Albert.
Measuring the universe.

Bibliography: p.187.
Includes index.
1. Astronomy—History. 2. Cosmological distances.
I. Title.
QB15.V33 1985 523.1 84-16397
ISBN 0-226-84881-7

Contents

Acknowledgments

The following story began as a few chapters in a projected, larger study on seventeenth-century telescopic astronomy, which remains still to be written. In the course of this project I incurred many debts.

Rice University supported a major portion of the research for this book through a sabbatical leave and a number of small research and travel grants over the last several years. A Research Fellowship from the National Endowment for the Humanities made a semester-long research trip to European libraries and archives possible.

Salomon Bochner of the Mathematics Department at Rice University was a constant inspiration to me while I was engaged in this project. Many of the ideas found on the following pages were first tried out on him during our long conversations. His passing is a great personal loss to me.

For the medieval and especially the ancient espisodes of this story I have relied heavily on the researches of Bernard Goldstein and Noel Swerdlow. Goldstein commented on several papers connected with this study and gave me many helpful comments on an early draft. Swerdlow read the entire manuscript and saved me from numerous errors. Curtis Wilson helped me reach a better understanding of Jeremiah Horrocks and pointed out a number of errors in other parts of the manuscript as well. Owen Gingerich read the final manuscript and made a number of useful suggestions. I also thank Wilbur Applebaum, the late M. L. Righini Bonelli, A. Rupert Hall, Marie Boas Hall, Michael Hoskin, Gerald Toomer, and Robert Westman for their answers to my questions and their comments on papers and drafts. Charles Garside, Jr., and Martin Wiener, my colleagues in the History Department of Rice University, read earlier drafts of this book and made numerous useful suggestions as to how I might improve the style and the argument. To all these colleagues I am deeply grateful. I bear sole responsibility for any remaining imperfections.

During the research for this book I received the help of many people. The hospitality and help of Maria Luisa Righini Bonelli of the Istituto e Museo di Storia della Scienza in Florence made my research there a joy. Her death is a great loss to the profession. Mara Miniatti of the museum staff helped me decipher several documents. Margaret Flood of the British Library went out of her way to help me obtain photocopies and microfilms. Mme de Narbonne and her staff at the Observatoire de Paris were very patient with me during the hot summer of 1983. I also wish to thank the staff of the library of the Académie des Sciences in Paris, the University

Library in Cambridge, the Humanities Research Center of the University of Texas at Austin, the Bibliothèque Nationale in Paris, the Staatsbibliothek in Munich, and the Universitätsbibliothek in Erlangen. I especially wish to thank the staff of the Fondren Library of Rice University for their patience and cooperation.

Finally, it is customary at this point for authors to acknowledge the moral support their spouses have given them. In my case, however, this support has gone far beyond mere moral support: Jo Nelson made it possible for me to take leave for a semester to write the final version of this book. I am deeply grateful for Jo's help and understanding, and I affectionately dedicate this book to her.

* 1 *
Shared Expectations

The recent *Pioneer* and *Voyager* missions to Jupiter and Saturn have once again demonstrated in spectacular fashion what impact a new technology can have on the body of scientific knowledge. All the existing information about Jupiter and Saturn, so slowly and painstakingly gathered in three and a half centuries of telescopic astronomy, was made obsolete almost in one stroke. Whereas astronomers traditionally had to wait for special observing opportunities, such as favorable oppositions or transits, to force nature to give up one small, precious morsel of information (and were often foiled by the weather), the data sent back by one planetary mission are usually so copious that only a tiny fraction of them can be examined reasonably quickly, and a significant portion may never be analyzed at all.

If the new space technology has had such an effect on planetary astronomy in the past two decades, then we would expect the new optical technology, the telescope, to have had an analogous impact on astronomy in the seventeenth century. One of the aims of this study is to reveal a subtle but important aspect of that impact. There is a complicating factor, however. When the telescope was first used in astronomy in 1609–10, that science was anything but static: it was in the middle of a revolution that had begun with the publication of *De Revolutionibus Orbium Coelestium* by Nicholas Copernicus in 1543. Although no historian would deny that the new instrument changed the course of this revolution, it proved to be anything but easy to document this change beyond the effects of the spectacular initial discoveries made by Galileo between 1609 and 1611.

The telescope revealed that Jupiter has moons and that Venus moves around the Sun, so there were at least three centers of motion (Earth, Sun, and Jupiter) in the cosmos. It also showed that the Moon is much like the Earth and that the Sun has blemishes, so it was much easier to think of the imperfect Earth as a heavenly body. After these initial revelations, however, what else did the telescope contribute to the debate between the proponents of Earth-centered and Sun-centered astronomy? It has been easier to show the instrument's impact on areas outside astronomy, such as literature, than to show how it affected the so-called Copernican Revolution after Galileo.[1]

To look for the impact of the telescope primarily in the realm of theory choice—Ptolemy or Tycho Brahe versus Copernicus—is misguided. Deep as the Great Debate cut through the astronomical community, it did not entirely separate the two camps from each other. At a fundamental level conservative and progressive astronomers still shared many traditional notions. It is one of these—shared expectations about sizes and distances—that we will examine in this study.

In 1631 Pierre Gassendi was the first astronomer ever to observe a transit of Mercury. This inferior conjunction was predicted by all three world systems, so that it could have no bearing on the Great Debate, but it was considered important for other reasons—the main one being an accurate determination of Mercury's size. Gassendi's record of the observation shows that he found it very difficult to believe his eyes: Mercury could hardly be as small as the dot he saw on the Sun. Others expressed the same surprise at Mercury's "entirely paradoxical smallness."[2] Obviously, astronomers had preconceived notions about the apparent sizes of the planets and fixed stars; notions that, as it turns out, led back a millennium and a half to Ptolemy. Gassendi had been taught a traditional, pretelescopic scheme of sizes and distances that was the common heritage of all educated Europeans and that had changed surprisingly little at the hands of Copernicus and Tycho Brahe. In the course of the seventeenth century, this scheme was replaced by the one with which we are familiar today. This change came about largely because of the telescope.

The telescope thus changed some very fundamental cosmological notions. Human societies usually have a cosmology and cosmogony that locate them in space and time. Such accounts are central to cultural identities. In Christian Europe this account was a mixture of the biblical creation story and late Greek cosmology received through Moslem intermediaries. A knowledge of this cosmogony and cosmology is central to an understanding of Christian Europe from the high Middle Ages to the seventeenth century. Although the qualitative aspect of medieval Christian cosmology has been elaborated in a number of scholarly and popular accounts, its quantitative aspect—the dimensions of the cosmos and all the bodies contained in it—has been ignored by all but the specialists in ancient and medieval astronomy. Yet, there is nothing particularly esoteric about this quantitative dimension of medieval cosmology. University students routinely learned it as part of the quadrivium, and it was therefore an integral part of the mental picture of the cosmos shared by educated men and women.

The scheme of sizes and distances that was welded into the heads of European students was not, however, the product of science as we know it. Rather, it was a mélange of philosophical notions, such as Aristotle's argument that the cosmos has no empty spaces; unverifiable naked-eye estimates, such as Ptolemy's statement that Mercury's apparent diameter is one-fifteenth of the Sun's diameter; and geometrical methods that produced spuriously precise results that had, in retrospect, more to do with numerology than with science. Its longevity resulted from the fact that it could not be falsified.

In this book we will follow this long tradition of cosmic dimensions from its beginning in antiquity to the seventeenth century. Then we will see how its tenacious hold on the minds of astronomers was slowly weakened by telescopic evidence in the first half of the seventeenth century. The telescope made it possible to subject this traditional scheme of sizes and distances to scientific scrutiny. With this instrument Gassendi could falsify Ptolemy's estimate of the apparent diameter of Mercury; indeed, he was almost forced to do so against his own will! In other words, with the telescope this quantitative aspect of cosmology began to change from a static philosophy to a progressive science.

If in the seventeenth century the telescope allowed a new consensus based on direct measurements to emerge for the apparent sizes of the planets, the all-important distance between the Sun and the Earth, the basis of all absolute distances, remained elusive. The usual view that the "astronomical unit" was first accurately determined by simultaneous measurements of the position of Mars in Cayenne and Europe in 1672 must now be doubted. To be sure, the length of this unit was progressively constrained within certain limits by factors indirectly related to telescopic measurements, but the length of the astronomical unit (which determined the entire new scheme of sizes and distances) that became accepted around the turn of the eighteenth century was a convenient estimate wrapped in the cloak of authority. Not until the expeditions to observe the transits of Venus in 1761 and 1769 can we say that solar parallax, and therefore solar distance, was measured directly.

The Venus expeditions of the eighteenth century opened a new age of international scientific cooperation, and they deserve special treatment on their own terms. In his important book *The Transits of Venus: A Study of Eighteenth-Century Science* (1959), Harry Woolf has described this episode in the history of science brilliantly. Therefore, the present book can end at the beginning of the eighteenth century, when the dimensions of the solar system and the sizes of the bodies in it, so familiar to us today, had come into rough focus. By that time Copernican astronomy, supported by a new physics, had won the Great Debate. The new astronomy and celestial physics as found in Isaac Newton's *Mathematical Principles of Natural Philosophy* (1687), however, dealt with a solar system that was very different from the one proposed by Copernicus in 1543. Besides adding no fewer than nine moons (four of Jupiter and five of Saturn), the telescope had expanded the solar system by an order of magnitude, and in doing so it had magnified the Sun to a size commensurate with its new central role and had tremendously enlarged the outer planets. Indeed, it had helped make the universe itself almost inconceivably large. The new cosmic dimensions, learned by all educated men and women, were one of the wonders of the age.

* 2 *
The Beginnings: Aristarchus and Hipparchus

Astronomical theory, as we know it, began with the Greeks. Their aim was to create an explanatory model of the cosmos that could also predict the positions of the Sun, Moon, and planets. From an early point, questions about the relative sizes of the parts of this cosmos and its overall dimensions—how large and how far away were the heavenly bodies—were an integral part of this quest. However, not until late antiquity, ca. A.D. 150, was a fully elaborated, coherent, quantitative picture of the cosmos finally achieved by Ptolemy. Yet Ptolemy built to a considerable extent on the foundation laid by his predecessors: Aristarchus of Samos (ca. 310–230 B.C.), Eratosthenes of Cyrene (ca. 276–195 B.C.), and Hipparchus of Nicaea (ca. 135 B.C.). These three men used the new tool of geometry to answer questions about the sizes and distances of the Earth and heavenly bodies. Our story must, therefore, start with them.

Tradition ascribes to Anaximenes of Miletus (sixth century B.C.) the notion that the Earth was a cylinder three times as wide as high, and that it was surrounded by three concentric rings carrying the fixed stars, the Moon, and the Sun. These rings were nine, eighteen, and twenty-seven times the diameter of the Earth, respectively.[1] Such numerological notions were perhaps precursors of spherical astronomy, of which Eudoxus of Cnidus (ca. 375 B.C.) is usually considered the originator. Eudoxus developed actual geometric models that could begin to account for the motions of the planets, if not to predict their positions. These models were refined by his younger contemporary, Calippus of Cyzicus (ca. 330 B.C.), and then linked together in a mechanical system of nesting spheres by Aristotle (384–322 B.C.). This system of a spherical Earth surrounded by concentric spheres of water, air, fire, and then the spheres of the heavenly bodies remained the basis for cosmology and physics for the next two millennia. But what was the order of the heavenly bodies and what were the dimensions of this cosmos?

In a geocentric, spherical cosmos the basis for all measurements, the astronomical unit, was the radius of the Earth. In the third century B.C., Eratosthenes of Cyrene gave the first geometric demonstration of how the size of the Earth could be determined. One of the postulates of spherical astronomy is that the Earth can be considered a mere point in relation to the spheres of the heavenly bodies. From this it follows that the Sun's rays striking the Earth are parallel, even at locations far removed from each other. Working in Alexandria, Eratosthenes assumed that Alexandria and

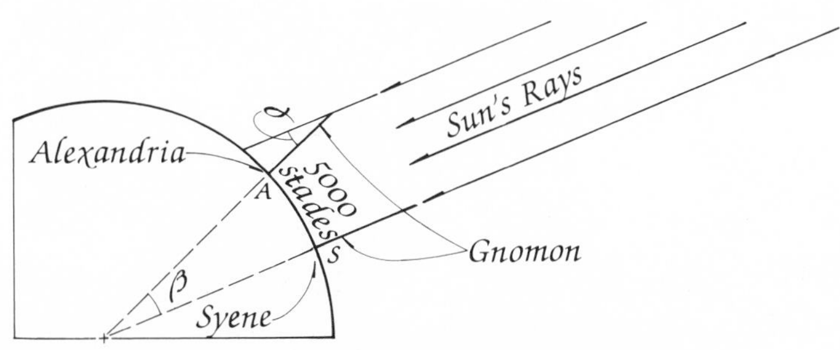

FIG. 1. Eratosthenes' demonstration of the size of the Earth.
$\angle \alpha$, the shadow at Alexandria, is equal to $\angle \beta$, the angle subtended at the Earth's center by the arc *AS*. Since $\angle \alpha = \angle \beta = 1/50$ of a circle, and $AS = 5000$ stades, the circumference of the Earth is 50×5000 stades, or 250,000 stades.

Syene (modern Aswân) lay on the same meridian, and he estimated the distance between them to be 5,000 stades (a stade being about a tenth of our mile). When the noonday Sun cast no shadow at Syene, its rays still cast a shadow in Alexandria (see fig. 1). By means of a gnomon (a vertical stick), Eratosthenes estimated the shadow angle to be one-fiftieth of a circle, and from this it followed that the Earth's circumference was $50 \times 5{,}000$, or 250,000 stades. A slight adjustment to 252,000 stades resulted in the length of a degree of 700 stades. The Earth's radius was thus slightly over 40,000 stades.[2]

Although Eratosthenes' procedure was a geometrical demonstration using only very approximate "measurements," his figure for the Earth's circumference gained great authority. Since we do not know the precise length of the stade he used, it is fruitless to speculate on the "accuracy" of his result. Suffice it to say that beginning with Eratosthenes the size of the Earth was known to the right order of magnitude.[3]

If it was a straightforward procedure to determine, at least roughly, the basic astronomical unit, it was extremely difficult to measure the distances to the heavenly bodies. Whereas the fairly obvious differences in the elevations of heavenly bodies with changes in latitude must have given the educated Greek traveler at least some idea of the Earth's size even without Eratosthenes' demonstration, no obvious phenomena would steer the mind toward any appreciation of the distances and absolute sizes of the heavenly bodies. It is thus quite possible that Anaxagoras of Clazomenae (ca. 445 B.C.) really did hold the opinion, ascribed to him by Diogenes Laertius, that the Sun was a fiery mass of ore the size of the Peloponnesus.[4] The first known attempt to determine the sizes and distances of any heavenly bodies by means of a geometrical demonstration was made by Eratosthenes' older contemporary, Aristarchus, remembered mainly for his heliocentric theory. While any work that Aristarchus may have written on Sun-centered astronomy has been lost, his little treatise *On the Sizes*

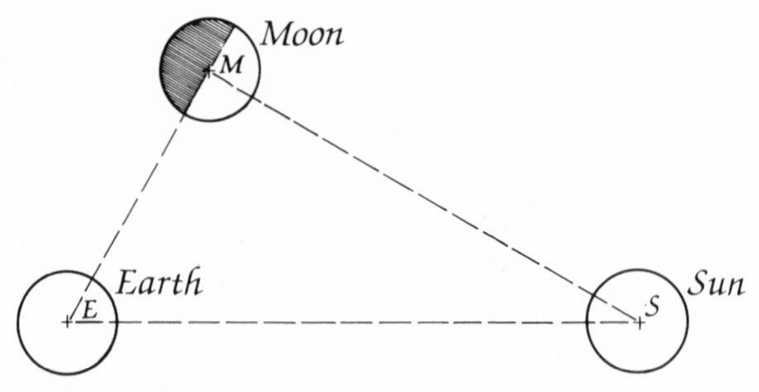

FIG. 2. Aristarchus's method of lunar dichotomy.
ME bisects the Moon's disk; $\angle EMS = 90°$; $\angle MES = 87°$. Aristarchus showed that $18{:}1 < ES{:}EM < 20{:}1$.

and Distances of the Sun and Moon has survived.[5] In it Aristarchus determined the ratios of the Sun's and Moon's distances from the Earth, as well as the ratios of their sizes to the size of the Earth.

Seen from a modern vantage point, *On the Sizes and Distances of the Sun and Moon* is a geometric success but a scientific failure. It shows at once the strength and the weakness of Greek geometry. Aristarchus developed two brilliantly imaginative geometric procedures—"lunar dichotomy" and the eclipse diagram—which were not discredited for two millennia. They are perfectly rigorous geometrical methods, but their results depend on an accuracy of measurement that is impossible to achieve, even today. His results (one of which became canonical) were therefore in error by an order of magnitude.

Not surprisingly, Aristarchus's little treatise is constructed on the model of Euclid's *Elements*. Necessary assumptions, such as that the Moon shines with reflected light from the Sun, and "measurements" are introduced as "hypotheses" at the very beginning.[6] The results are then stated as propositions, followed by the demonstrations of these results.

Aristarchus's most celebrated demonstration is the bisected Moon, or lunar dichotomy. Twice during each cycle of waxing and waning, at First Quarter and Last Quarter, the Moon appears to be exactly half illuminated or "dichotomized." At that moment, when the terminator (the dividing line between the illuminated and unilluminated part) exactly bisects the Moon's disk, as in figure 2, $\angle EMS$ is 90°. If at this precise moment we measure $\angle MES$, the angular separation between Moon and Sun, all angles in triangle *EMS* will be known, and we can therefore determine the ratios of the sides.

Aristarchus assigned $\angle MES$ the value "less than a quadrant by a thirtieth of a quadrant," that is, 87°.[7] This is, in all likelihood, a convenient value that we should not take as an accurate measurement.[8] At this point we would simply look up the secant of 87° and arrive at the ratio $ES{:}EM = 19.1{:}1$, but Aristarchus did not have trigonometric functions and ta-

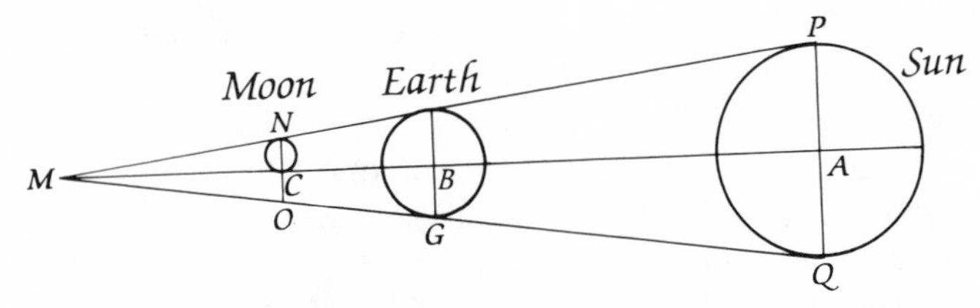

FIG. 3. Aristarchus's use of the eclipse diagram to determine the actual sizes of the Sun and Moon

bles, and he had to use a laborious procedure to show that the ratio of *ES* to *EM* is greater than 18:1 but less than 20:1.[9]

Now, in fact, the Sun is about 390 times as far removed from the Earth as is the Moon. Aristarchus's geometry was impeccable, but his method proved to be impractical. Even if he would have tried to measure his numerical data accurately, he would have found that determining the exact moment of dichotomy and then measuring the angular separation of the two luminaries accurately is a hopeless task. Even a small error in the time of dichotomy leads to a significant error in $\angle MES$, itself difficult to measure accurately. And from the diagram it is clear that a small error in $\angle MES$ will produce a proportionally much larger error in the all-important $\angle MSE$, the angle subtended by the Earth-Moon distance for an observer on the Sun. The inaccuracies are thus compounded. To illustrate this, $\angle MES$ should be about 89°50′, not 87°. The discrepancy of 2°50′ means an error of *a factor of eighteen* in the desired ratio of distances.[10]

Having established the ratio of the two distances, Aristarchus went on to establish the ratio of the actual diameters of the Sun and Moon. For this he needed to know the ratio of apparent diameters, but this was easy. As Aristarchus put it:[11]

> When the sun is totally eclipsed, the sun and the moon are then comprehended by one and the same cone which has its vertex at our eye.

In other words, he made the apparent diameters of the Sun and Moon equal, as is evident during a solar eclipse.[12] From this it follows directly that their actual diameters are in the same proportion as their distances from the Earth.[13]

So far, all measures determined were relative. To determine the *absolute* sizes of the Sun and Moon, Aristarchus introduced a second geometric procedure: the eclipse diagram, a method that, when fully developed by Hipparchus and Ptolemy, was to be the centerpiece of all determinations of absolute celestial distances until the seventeenth century.

In figure 3 the Moon is entirely immersed in the shadow cone of the Earth during a lunar eclipse. Aristarchus stated that *NO*, the width of the Earth's shadow cone at the Moon's distance, is equal to twice NC, the

Moon's diameter.[14] He had presumably determined this measure at least roughly from an analysis of lunar eclipses. He also knew from the lunar dichotomy procedure that $18/1 < AB/BC < 20/1$. He argued as follows:[15]

We know that $AB/BC > 18/1$, and that $NO = 2NC$.
Therefore, $NO/PQ < 1/9$, and $AM/MC > 9/1$.
From this, $AM/AC < 9/8$. (1)
Also, since $AB/BC > 18/1$, $AC/AB < 19/18$. (2)
Combining (1) and (2), $AM/AB < (9 \times 19)/(8 \times 18)$,
or $AM/AB < 19/16$.
And thus, $AM/BM > 19/3$ and $AQ/BG > 19/3$.

Using the same procedure for $AB/BC < 20/1$, Aristarchus showed that $AQ/BG < 43/6$.[16] The diameter of the Sun was thus more than $19/3$ but less than $43/6$ times as great as the diameter of the Earth. And using the ratio for the absolute sizes of the Sun and Moon found above, he showed that the Earth was more than $108/43$ but less than $60/19$ times as great in diameter as the Moon.[17] In sum, expressing Aristarchus's results in convenient mean figures, we can say that he determined the Sun to be 19 times as far from the Earth as is the Moon, and that he found the Sun to be about $6\frac{3}{4}$ times as large and the Moon about $1/3$ as large as the Earth.

From these absolute sizes of the Sun and Moon, their absolute distances can easily be calculated, provided we know their apparent sizes, each about $1/2°$. Here there was a problem, however. For although from a reference by his younger contemporary Archimedes (ca. 287–212 B.C.) we are certain that Aristarchus had actually measured the Moon's apparent diameter to be $1/720$th part of its orbit, that is, $1/2°$,[18] in *On the Sizes and Distances of the Sun and Moon* he stated that the Moon's apparent diameter was 2°.[19] He had used this exaggerated value to demonstrate that although the Sun is larger than the Moon and thus must illuminate more than a lunar hemisphere, this excess is imperceptible to the observer on Earth and therefore negligible.[20] Presumably Aristarchus chose such a large value for the apparent diameter to make the demonstration more convincing: if the excess illumination is not perceptible in a globe with an apparent diameter of 2°, it will certainly be imperceptible in a globe that appears only a fourth as big.[21] To us this demonstration may seem superfluous and even pedantic; to Aristarchus it was important because he was writing a formal geometric treatise.

Knowing that (in rough numbers) the Moon's actual diameter is about 0.35 Earth diameters, or 0.7 earth radii (e.r.), let us take 2° as the Moon's apparent diameter, which is 1/180th part of the circumference of its orbit. The circumference of the Moon's orbit is, therefore, 180×0.7, or 126 e.r., and its radius, which is the distance from the Earth to the Moon, is $126/2\pi$, or 20 e.r. The distance to the Sun is then, roughly, 19×20, or 380 e.r. If, however, we use the more reasonable value of $1/2°$ for the Moon's apparent diameter (a value which, as we have seen, Aristarchus himself actually measured), the distances of the Moon and Sun come out to be roughly 80 and 1,520 e.r., respectively.

Aristarchus did not calculate these absolute distances, however! After a determination of the ratio of volumes of the Moon and Earth, the tract ends abruptly. Perhaps standards of rigor would not allow the introduction of a different and more reasonable value for the Moon's apparent diameter at this point. Perhaps Aristarchus was not concerned with absolute distances. We can only speculate.

In this pioneering tract, Aristarchus presented the world with two geometric constructions for determining sizes and distances in the heavens: lunar dichotomy and the eclipse diagram. Whereas in his work the two methods were linked (the results of the former being used in the latter), in the works of his successors the two were separate. The eclipse diagram, in a somewhat more elaborate form, came to play the leading role in distance determinations, while lunar dichotomy was cast in a minor role, until it briefly came into the spotlight in the seventeenth century. It was not until late in the seventeenth century that both methods finally became obsolete.

Aristarchus's treatise, however, addressed only the problem of the sizes and distances of the two great luminaries. No comparable geometric methods, however inadequate by our standards, were at hand for determining the sizes and distances of the other heavenly bodies. Indeed, even the order of the planets was a question without a definite answer. The Moon had to be the lowest celestial body since it eclipses all others. Since the fixed stars were the immobile reference frame against which all motions were plotted, they were naturally put in the outermost sphere. The sensible order was to arrange the bodies according to their speeds with respect to the fixed stars, beginning at the top with Saturn, which takes 30 years to return to the same place in the zodiac, and ending with the Moon, which takes about 30 days for the same trip. Jupiter and Mars followed Saturn in downward order, but between Mars and the Moon it was impossible to decide on the order of the Sun, Mercury, and Venus using this criterion, since all three take a year to traverse the zodiac. As we shall see, a consensus on the order of the planets came only when Ptolemy put his vast authority behind one particular order.[22]

In the meantime, any complete system of planetary distances had to be based on a tentative order and derived from criteria other than measurements, criteria such as harmony and numerology. A good example is a system ascribed to Archimedes by the Christian bishop Hippolytus (third century A.D.). Although by the time of Hippolytus the text and numbers had become garbled, it appears that Archimedes used two planetary orders: Moon-Sun-Venus-Mercury-Mars-Jupiter-Saturn, and Moon-Venus-Mercury-Sun-Mars-Jupiter-Saturn. The distances themselves do not reveal any rationale based on measurements, and we are at a loss as to how Archimedes arrived at them. In this scheme he made the distance to the fixed stars, the radius of the cosmos, 248,264,780 stades, or, using 40,000 or 50,000 stades as the Earth's radius, 5,000 or 6,000 e.r.[23]

In his *Sand Reckoner*, which has survived, Archimedes was not so much concerned with technical astronomical or cosmological matters as with large orders of magnitude. Setting himself the task of showing that the number of grains of sand needed to fill the cosmos could be expressed in numbers, he chose convenient upper limits for cosmic distances. He made the distance to the Sun less than 10,000 e.r., and then, in Aristarchus's hypothetical heliocentric universe (the largest universe ever proposed up to that time), he assumed that the diameter of the Earth's orbit is as a point compared to the vast distance to the fixed stars. This distance was, then, at most 100,000,000 e.r.[24] As in his other estimate of cosmic distances, very little astronomy was involved, but Archimedes did describe in this tract how to measure the apparent diameter of the Sun by means of a *dioptra* equipped with a movable, vertical cylinder that could be slid into position to block out the Sun exactly. By this means he determined that the Sun's angular diameter lay between the limits $90°/200$ and $90°/164$, or between $27'$ and slightly less than $33'$.[25]

Although harmonic and numerological speculations about the sizes and distances of heavenly bodies were fairly common, Aristarchus's geometric treatment, based on what were at least potential measurements, remained unique for about a century. Hipparchus of Nicaea, working in Rhodes in the third quarter of the second century B.C., was the first astronomer known to use Aristarchus's methods in the service of technical astronomy. Except for one minor book, the works of this important astronomer have been lost, including one entitled *On Sizes and Distances*. In the last several decades, however, scholars have discovered much about Hipparchus's achievements, and we now have a reasonably good idea of how he improved on Aristarchus's approach to the problem of sizes and distances.[26]

Whereas all his Greek predecessors had been satisfied with geometrical models that could in principle account for the motions of heavenly bodies, Hipparchus set out to construct models with actual predictive value—geometrical models with parameters derived from actual observations. He had available to him Babylonian eclipse observations going back to the eighth century B.C., and with the help of these and his own observations he arrived at a good solar theory and a lunar theory that gave good results at the syzygies, that is, New Moon and Full Moon. With these two theories Hipparchus could now predict lunar eclipses, but solar eclipses were more difficult.[27]

While the Moon appears eclipsed to the same degree to all observers for whom it is above the horizon, a solar eclipse will be total only in a narrow band. Observers outside this area will see either a partial eclipse or no eclipse at all. In figure 4 the observer at *A* will see a total eclipse of the Sun. What is seen by the observer at *B* depends on the distance *AB* and the distances to the Moon and Sun, or, more conveniently, the angles subtended by *AB* at the Moon and Sun, $\angle AMB$ and $\angle ASB$. These parallax angles are at their maximum when for the observer at *A* the Moon and Sun are at

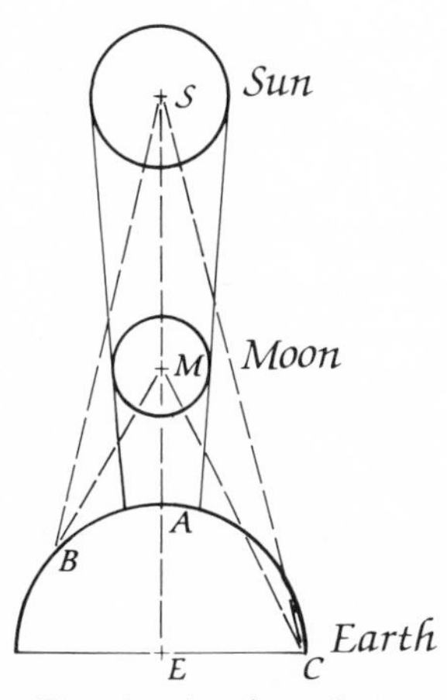

FIG. 4. A solar eclipse

the zenith while for the observer at C they are on the horizon. Angles AMC and ASC are the horizontal parallaxes of the Moon and Sun, fundamental parameters in eclipse astronomy. To predict solar eclipses, then, Hipparchus had to determine the distances or parallaxes of both the Sun and Moon. It was impossible to measure these quantities directly, since this would have required two observers at widely separated locations to measure the position of the Sun or Moon simultaneously. However, since Hipparchus was convinced that the Sun's parallax is very small, perhaps imperceptible, he made assumptions about solar parallax and then determined the much greater lunar parallax.

Hipparchus's first approach was to use information from a previous solar eclipse (recently identified as that of 14 March 189 B.C.)[28] during which the Sun was totally eclipsed near the Hellespont and four-fifths eclipsed at Alexandria. If one assumed that the Sun's parallax between these locations was zero, then the difference between total and partial eclipse, one-fifth the Sun's angular diameter, was due entirely to the parallax of the Moon between the Hellespont and Alexandria. Hipparchus had measured the Moon's apparent diameter at mean distance to be 1/650th of its orbit, that is, $360°/650$ or $33\frac{1}{4}'$. Assuming the Sun's and Moon's apparent diameters to be equal, one-fifth of the Sun's apparent diameter was therefore about $6\frac{2}{3}'$, and that was the lunar parallax between the Hellespont and Alexandria. He also knew the latitudes of the Hellespont (41°) and Alexandria (31°) as well as the Moon's elevations at these two locations. This was enough to allow him to calculate the Moon's distance, although we do not know his exact procedure. Hipparchus's results were 71 and 83 e.r. for the Moon's least and greatest distances.[29]

In his second approximation Hipparchus improved on the eclipse diagram of Aristarchus. He did not use lunar dichotomy to establish the ratio

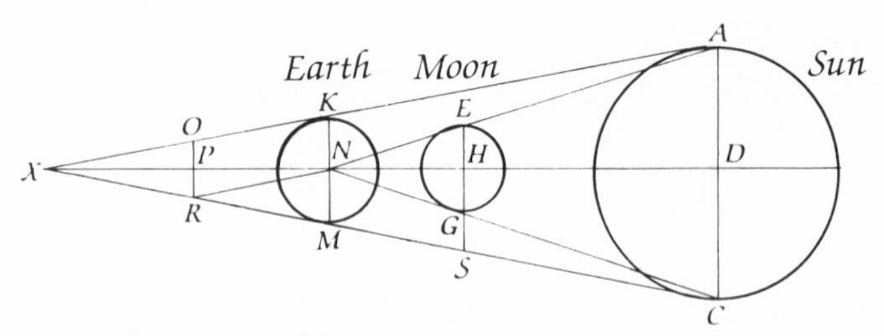

FIG. 5. Hipparchus's use of the eclipse diagram to determine the lunar distance

of solar to lunar distances, however, and therefore his initial information was limited to three measurements:[30]

1. At mean distance the Moon's apparent diameter is 1/650th part of its orbit.
2. At mean distance the Moon's apparent diameter is equal to the Sun's.
3. At mean distance the width of the Earth's shadow cone is 2½ times the lunar diameter.

This was not enough to find the distances of the Sun and Moon, and Hipparchus therefore *assumed* a horizontal solar parallax of 7′, the smallest perceptible, or perhaps the greatest imperceptible, parallax. This was the equivalent of assuming a solar distance of 490 e.r. (1/490 = sin 7′). We know from Ptolemy that Hipparchus's method of finding the Moon's distance was more or less as follows.[31]

In figure 5 (not drawn to scale), we know that the apparent diameters of the Sun and Moon, $\angle ENG$ and $\angle ANC$, are equal, and that $\angle ENG$ is 360°/650. The width of the shadow cone at the Moon's distance, *OR*, is 2½ times *EG*, the Moon's diameter; the lunar distances *NP* and *NH* are equal; and the solar distance *ND* is 490 times *NM*, the Earth's radius.

Consider first the trapezoid *RPHS* (fig. 5a), in which $PR \parallel NM \parallel HS$, and $NH = NP$.

$PR + HS = 2NM$, so that $HS = 2NM - PR$. (1)
Also, $PR = 2\tfrac{1}{2}\,HG$. (2)
Substituting (2) into (1), $HS = 2NM - 2\tfrac{1}{2}\,HG$. (3)
And since $GS = HS - HG$,
$GS = 2NM - 2\tfrac{1}{2}\,HG - HG = 2NM - 3\tfrac{1}{2}\,HG$. (4)

Consider now trapezoid *NMCD* (fig. 5b), in which $NM \parallel HS \parallel DC$.

In triangles *NMC* and *SGC*, $NM/GS = NC/GC$. (5)
In triangles *NDC* and *NHG*, $NC/GC = ND/HD$. (6)
Therefore, $NM/GS = ND/HD$. (7)
Replacing *HD* by $ND - NH$,
$NM/GS = ND/(ND - NH)$. (8)
Solving for *NH*, the unknown,
$NH = ND(NM - GS)/NM$. (9)

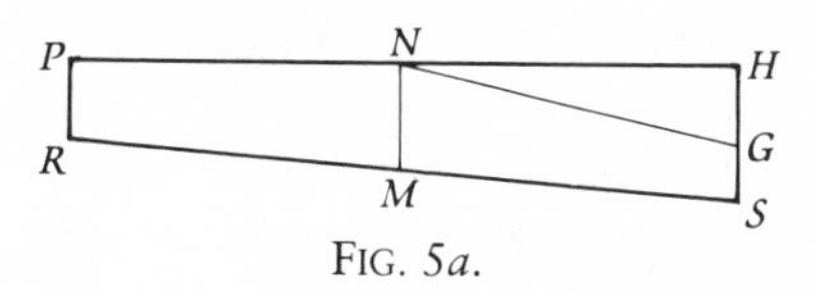

FIG. 5*a*.

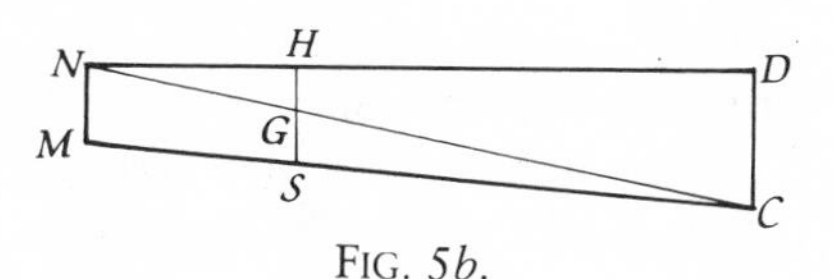

FIG. 5*b*.

Combining (4) and (9),

$$NH = ND(3\tfrac{1}{2}\,HG - NM)/NM. \qquad (10)$$

But HG, the Moon's radius, is 1/1,300th of the Moon's entire orbit, whose radius is NH, so that $HG = 2\pi NH/1{,}300$.

Therefore,

$$NH = ND(7\pi NH/1{,}300 - NM)/NM. \qquad (11)$$

But NM is the unit in which we want to express NH, so that we can write $NH = ND(7\pi NH/1{,}300 - 1)$, and, solving for NH,

$$NH = ND/(7\pi ND/1{,}300 - 1). \qquad (12)$$

Since we know that $ND = 490$ e.r., $NH = 67.2$ e.r. Hipparchus found a slightly larger value: $NH = 67\tfrac{1}{3}$ e.r.[32]

Hipparchus had used the smallest possible solar distance, and the geometry of the diagram dictates that this would yield the greatest possible lunar distance. Because he had used the parameters for the Moon's mean distance, he had therefore found the largest possible value for the Moon's mean distance. He also concluded that the smallest possible mean lunar distance was 59 e.r., which corresponds to letting ND go to infinity in equation (12). It is not known whether Hipparchus considered the effects on NH of varying the other two parameters, the apparent diameter of the Moon and the width of the shadow cone. Small variations in those parameters can substantially alter the results of a procedure such as this. We shall say more about this in the discussion of Ptolemy's procedure.

Although Hipparchus made no headway in determining the Sun's distance independently, his assumption of solar parallaxes and distances did lead him to roughly converging values of lunar parallaxes which were more important in solar eclipse calculations. If he arrived at a more comprehensive scheme of the sizes and distances of all heavenly bodies, which is unlikely, this has not been preserved. We do know from Ptolemy, however, that Hipparchus had some ideas about the apparent diameters of at least some of the other heavenly bodies:[33]

> Hipparchus said that the apparent diameter of the Sun is 30 times as great as that of the smallest star, and that the apparent diameter of Venus, which appears to be the largest star, is about a tenth the apparent diameter of the Sun.

These were, of course, the merest of comparative estimates, and we do not know if Hipparchus himself put much faith in them. Only with Ptolemy, three centuries later, were such estimates to become important. It should be noted, in the meantime, that Hipparchus and his successors up to Galileo did not distinguish between brightness and size.

Hipparchus's achievements appear to have had little impact on his immediate successors. Posidonius (ca. 140–50 B.C.), who spent the last part of his life in Rhodes where Hipparchus had worked, ignored the astronomer's results (if he knew of them) and used a simplified method to estimate the lunar distance. Assuming that the Earth's shadow was a cylinder instead of a cone, and that its width was twice the Moon's diameter, he concluded that the Moon's actual diameter was half the Earth's, or 40,000 stades. Using a measurement of the Sun's and Moon's apparent diameters of 1/750th part of their orbits, he concluded that the circumference of the lunar orbit was 750 × 40,000 stades or 30,000,000 stades, and its radius, the lunar distance, 5,000,000 stades. This corresponds to 125 e.r. for the lunar distance.[34]

For the distance of the Sun, Posidonius resorted to mere speculation, assuming in one case that the Sun's orbit was 10,000 times the Earth's circumference, and in another that the Sun and Moon moved with the same linear velocities. The first assumption leads to a solar distance of 10,000 e.r. while the second produces a value of 1,625 e.r.[35]

Posidonius's efforts were clearly mere guesses. Others indulged in similar unfounded speculations, and by the turn of the Christian era the philosophical literature contained a number of different guesses besides the geometrically derived results of Aristarchus and Hipparchus. Despite all its shortcomings, however, the geometric approach of Aristarchus, augmented by Hipparchus, was the only solid basis on which astronomers could build. A complete system of astronomy was impossible without lunar and solar parallaxes, and for these the eclipse diagram offered the only hope. When, in the second century A.D., Ptolemy actually managed to construct a system of astronomy that could not only account for the Moon and Sun but also for the planets, an important by-product was a complete and coherent system of cosmic dimensions. The geometric approach of his two great predecessors was central in this achievement.

* 3 *
Ptolemy

From the second to the sixteenth century, astronomy was a commentary on Ptolemy. No man ever wielded posthumously such a pervasive and long-lived authority in astronomy, and it is to be doubted that anyone ever will again. Ptolemy's work superseded the efforts of all his predecessors—surely one of the main reasons why so few of their works have survived—and it defined the astronomical problems for his successors, at least until the time of Tycho Brahe and Johannes Kepler. At the center of the Ptolemaic tradition stands his *Syntaxis Mathematica*, known in the Latin West by the title *Almagest*,[1] a magnificent treatise on mathematical astronomy containing predictive models for the Moon, Sun, and planets. Among his other works were major treatises on astrology, geography, and optics, as well as a small, speculative cosmological work entitled *Planetary Hypotheses*. His cosmic dimensions, derived in part from the mathematical models of the *Almagest*, are presented in his *Planetary Hypotheses*.

Ptolemy's scheme of sizes and distances, the first complete and coherent quantitative cosmology in the Western tradition, is often referred to as the "Ptolemaic System," a term adopted here. The historical significance of this system cannot be sufficiently stressed. At the center of the Ptolemaic achievement in astronomy lay a tension between mathematical, predictive astronomy and physical, explanatory cosmology, because the former employed devices, such as the eccentric and equant, which violated the latter's basic tenets of perfect circular motion. The Ptolemaic System, however, provided a bridge between these two incommensurate approaches: while, on the one hand, it welded the unconnected planetary models of the *Almagest* into a system, on the other hand, it provided the qualitative cosmological system derived from Aristotle with a much needed quantitative dimension. For this reason it attained a very powerful grip on the minds of people in the Middle Ages and during the Renaissance.

In the *Almagest* Ptolemy was concerned with predicting the positions of heavenly bodies. In the cases of the planets, sizes and distances played no role in these calculations, but the Moon has a noticeable parallax, and lunar observations had to be corrected to reduce them to the Earth's center. Moreover, in the predictions of solar eclipses the parallaxes of the Moon and perhaps of the Sun could not be ignored. Ptolemy therefore had to determine the parallaxes of the Moon and Sun.

Since, as we know, the Moon's motion is significantly influenced by the Sun, it is difficult to reduce to a mathematical model. Lunar theory was the bane of astronomers from antiquity to the eighteenth century. With a simple epicyclic model, such as Hipparchus had already developed, Ptolemy could give a satisfactory account of lunar positions at or near the syzygies, but this model was woefully inadequate at the intermediate positions, the quadratures. He therefore constructed a more complicated model with a movable eccentric and an epicycle, which accounted for the Moon's positions at the quadratures and reduced to the simpler model at the syzygies.[2] The dimensions of this model were such that the maximum and minimum relative distances of the Moon at the syzygies and quadratures were:[3]

Maximum distance at syzygy	65.25 parts
Minimum distance at syzygy	54.75 parts
Maximum distance at quadrature	44.62 parts
Minimum distance at quadrature	34.12 parts

In order to turn these relative distances into absolute ones, Ptolemy needed to measure one absolute lunar distance or parallax. He did this by comparing a measured position of the Moon with its calculated position at that time. He found what amounts to a horizontal lunar parallax of 1°26′ at the time of the observation,[4] a value that is much too large. The error can be ascribed to a combination of factors, such as failure to correct the observation for the effects of atmospheric refraction, instrument errors, and inaccuracies in his theory. Ptolemy's lunar parallax corresponds to an absolute lunar distance of 39.75 e.r., while his lunar model gave 40.25 parts for the relative distance at this position. A factor of 39.75/40.25, or about 59/60, would therefore transform all relative lunar distances into absolute ones. At its farthest remove from the Earth, which occurred at syzygy, the Moon's distance was, thus, 64⅙ e.r., and at its closest approach, which occurred at quadrature, its distance was 33.55 e.r.[5]

Ptolemy could now draw up a table of lunar parallaxes. It is apparent, however, that whereas the distances and parallaxes at the syzygies are fairly close to the modern values, those at the quadratures are very far removed from them. If Ptolemy had no way of knowing what astronomers a millennium or two after his death would conclude about lunar parallaxes and distances, he certainly knew his results predicted that at the least distance at quadrature the Moon's apparent diameter should be nearly twice as large as at the greatest distance at syzygy. This is manifestly not the case, as Ptolemy knew very well, but he did not comment on the discrepancy.[6]

Having solved the problem of lunar parallaxes and distances, Ptolemy turned his attention to solar parallax and distance. He wished to use the eclipse diagram of Aristarchus and Hipparchus, but to do so he needed to know the apparent diameters of the Sun and Moon, as well as the diameter of the shadow cone at the Moon's distance. From measurements with a

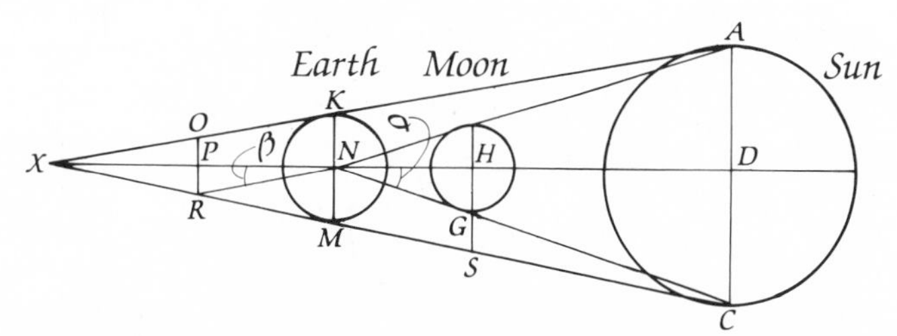

FIG. 6. Ptolemy's use of the eclipse diagram to determine the solar distance

dioptra he could only conclude that, even though his solar theory predicted a ratio of greatest to least solar distances of 62½ :57½, the Sun's apparent diameter did not vary, and that it was equal to the Moon's apparent diameter at the greatest lunar distance.[7] This last conclusion would rule out an annular solar eclipse, one in which the Moon's disk is centered on the Sun's disk but is too small to cover it completely.

Since the *dioptra* was not sufficiently accurate to provide Ptolemy with the measurements he needed, he resorted to an analysis of lunar eclipses. From a comparison of two partial lunar eclipses recorded by Babylonian astronomers in 620 and 522 B.C., he concluded that the Moon's apparent diameter at its greatest distance of 64⅙ e.r. was 31′20″, while the diameter of the shadow cone at this distance was 1°21′20″, or "very little less than 2⅗ times the diameter of the Moon."[8] The selection of these two eclipses so far in the past, coupled with our knowledge that estimating the magnitude of an eclipse is extremely difficult and that Ptolemy's procedure was sensitive to even small errors in these estimations, indicates that he was probably looking for particular values. We shall return to this problem.

Ptolemy now had all the ingredients necessary to use the eclipse diagram. But whereas Hipparchus had assumed a solar distance and derived a lunar distance, Ptolemy had determined the lunar distance by a different method and reversed Hipparchus's procedure in order to derive a solar distance.

In figure 6 the lunar distance NH is chosen as 64⅙ e.r., the Moon's greatest distance, at which its apparent diameter is 31′20″ and equal to the Sun's apparent diameter. Therefore $\angle\alpha = 15'40''$. Furthermore, the diameter of the shadow cone is 2⅗ times the lunar diameter, so that $\angle\beta = 2\frac{3}{5}\angle\alpha$. Consider triangle NHG in figure 6. We know that $NH = 64\frac{1}{6}NM$, or, setting NM equal to unity, $NH = 64\frac{1}{6}$, and $\angle\alpha = 15'40''$. We can, therefore, find HG:

$$HG = NH \tan 15'40''.$$

Ptolemy used chords and found $HG = 0.2925$.

We also know that $PR/HG = 2\frac{3}{5}$,

so that $PR = 0.7605$.

In trapezoid $RPHS$, $PR + HS = 2NM$, or
$PR + HS = 2$. (1)
But $PR + HG = 0.2925 + 0.7605 = 1.0530$. (2)
Subtracting (2) from (1), $HS - HG = GS = 0.947$. (3)
In triangle NMC, $NM/GS = NC/GC$. (4)
In triangle NDC, $NC/GC = ND/HD$. (5)
Combining (4) and (5), $NM/GS = ND/HD$, or
$ND/HD = 1/0.947$, and $HD = 0.947ND$.
But $HD = ND - NH$, where $NH = 64\frac{1}{6}$ e.r.,
and thus,
$0.947ND = ND - 64\frac{1}{6}$ e.r., or $ND = 1{,}210$ e.r.[9]

This was the procedure used and the answer obtained by Ptolemy. It is useful to express ND in general terms, using modern trigonometric functions.

$$ND = NH/\{(\angle\beta/\angle\alpha + 1)\, NH \sin\angle\alpha - 1\}, \qquad (6)$$

to show ND as a function of NH, $\angle\alpha$, and $\angle\beta/<\alpha$.

Ptolemy had thus managed to derive a solar distance from parameters more or less based on actual observations. His procedure needs some comments, however. First and foremost, whereas Hipparchus's procedure resulted in a relationship (ch. 1, eq. [12]) that is not sensitive to small variations in the parameters, Ptolemy's relationship (6) is extremely sensitive to such variations. If in (6) we make changes on the order of 1 percent in one parameter while keeping the others constant, the results are as follows:

1. If $\angle\alpha$ is changed from $15'40''$ to $15'30''$, ND becomes 1,545 e.r. If $\angle\alpha$ is $13'$ or less, ND is infinite.
2. If NH is changed from $64\frac{1}{6}$ e.r. to 65 e.r., ND becomes 979 e.r. If we make NH 61 e.r. or less, ND becomes infinite.
3. If we make the width of the shadow cone "very little less than $2\frac{3}{5}$," say, 2.57 times the Moon's apparent diameter, so that $\angle\beta/<\alpha = 2.57$, ND becomes 1,451 e.r. If $\angle\beta$ is $38'$ or less, ND becomes infinite.
4. If we make three simultaneous changes, all tending to make ND larger, say, $\angle\alpha = 15'30''$, $\angle\beta/<\alpha = 2.57$, and $NH = 63.5$ e.r., then ND becomes 2,872 e.r.[10]

Such variations of the parameters are an order of magnitude smaller than the margins of error of the procedure used by Ptolemy to determine them. Where Hipparchus's approach of assuming a solar parallax and calculating the corresponding lunar parallax was not particularly sensitive to small changes of the parameters, Ptolemy's method of calculating a solar distance was very sensitive to such variations. Although the methods of Hipparchus and Ptolemy are geometrically the same, Hipparchus's procedure was much better science.

Second, Ptolemy's solar distance of 1,210 e.r. is almost exactly 19 times as great as his lunar distance of $64\frac{1}{6}$ e.r. This was precisely the ratio arrived at by Aristarchus via the method of lunar dichotomy. Granted that this may very well have been sheer coincidence, the agreement nevertheless later served to strengthen the authority of Aristarchus's ratio and Ptolemy's solar distance. In view of the sensitivity of equation (6), it is not unreasonable to argue that Ptolemy knew beforehand, at least roughly, what he wanted his solar distance to be, and arranged, $\angle\alpha$ and $\angle\beta$, and perhaps also *NH*, accordingly. Although Otto Neugebauer warns us against drawing such conclusions,[11] others have not hesitated to do so. Willy Hartner wrote recently:[12]

> Ptolemy's elaborate report on his procedure, to find from two early Babylonian lunar eclipses an exact value for the Moon's [apparent] diameter (31′20″) without having recourse to the *dioptra* (which he says yields no reliable results), is a fairy-tale. What he actually did, was to start from Aristarchus' mean value [*ND* = 19*NH*], and to compute from it the appropriate value for [the Moon's apparent diameter]. He then makes us believe . . . that he derived his value for [the lunar diameter] in the mentioned nonsensical way and found from it the distance of the Sun to equal 19 times that of the Moon.

Because of procedures such as this, Robert Newton has recently gone so far as to accuse Ptolemy of fraud.[13] Kepler, who spent many years investigating this particular method, judged Ptolemy more kindly :[14]

> . . . if anyone seeks very carefully into the method which Ptolemy employed for establishing the distance of the Sun, he will very greatly praise the singular ingenuity of the demonstration; but he will pronounce those things which Ptolemy accepted as very suspect, as if provided for the purpose of demonstrating that which Ptolemy had taken from the ancients.

Last, the complicated procedure used by Ptolemy to establish the solar distance is, in retrospect, puzzling. Consider again figure 6:

$$\angle\alpha + \angle\beta = 180° - \angle RNC,$$

$$\text{and } \angle NRM + \angle NCM = 180° - \angle RNC.$$

$$\text{Thus, } \angle\alpha + \angle\beta = \angle NRM + \angle NCM.$$

But $\angle NRM$ is the lunar parallax and $\angle NCM$ is the solar parallax. Therefore, $\angle NRM = 15'40'' + 40'40'' - \csc^{-1} 64\frac{1}{6} = 2'45''$, corresponding to a solar distance of 1,250 e.r. Why did Hipparchus and Ptolemy, and their successors up to Kepler,[15] use this cumbersome method when the simpler one lay at hand?

*　　*　　*

Having obtained the absolute distances (and therefore the absolute sizes) of the Sun and the Moon, Ptolemy could predict solar eclipses and construct tables of solar and lunar parallaxes. He ignored the eccentricity of the Sun's motion, so that solar parallaxes could be read off simply against the Sun's zenith distances, varying from zero at the zenith to 2′51″ at the horizon. In the case of the Moon, eccentricity could not be ignored, therefore the tables required computations based on the Moon's position in its orbit. For the greatest distance at syzygy (64⅙ e.r.) the tables lead to a horizontal lunar parallax of 53′, while for the least distance at quadrature (33.55 e.r.) they yield a horizontal parallax of 1°42′.[16]

For predictive purposes Ptolemy had done all he needed to do; he did not treat the subject of the sizes and distances of the other heavenly bodies in the *Almagest*. He did, however, briefly raise the subject of the order of the heavenly bodies in the introduction to the theories of the planets. As mentioned in the previous chapter, in geocentric astronomy the order of the planets between the Moon and the fixed stars was a matter of convention. Ptolemy agreed with his predecessors on putting Saturn, Jupiter, and Mars below the fixed stars in that order. But what was the order between the Moon and Mars? He wrote:[17]

> . . . we see that the spheres of Venus and Mercury were placed below the spheres of the Sun by the ancient astronomers but were placed above the same by some later astronomers, since a passage of these planets across the Sun has never occurred. This ostensibly decisive reason seems to us not to be valid because there can be planets below the Sun without these having to lie throughout in a plane passing through the Sun and our eye. They can, rather, lie in another plane and for that reason effect no apparent passage across the Sun, just as at the times of conjunctions, when the Moon, whose path runs underneath the Sun, passes the Sun, usually no eclipses occur.

Ptolemy himself, however, could not propose more compelling criteria by which to decide the order of the Sun, Venus, and Mercury:[18]

> But since this question cannot be decided in any other way since none of these planets shows a perceptible parallax—the sole phenomenon that allows the distances to be determined—the order of the ancient astronomers seems to deserve greater credibility. This order, more naturally appropriate to the central position of the Sun, separates the planets which reach opposition from those which do not reach this position but, rather, always remain in the vicinity of the Sun. At any rate, this arrangement of these last planets below the Sun may not result in such a great proximity to the Earth that the approach could result in a perceptible parallax.

Ptolemy thus believed that Venus and Mercury were below the Sun, but that they were far enough above the Earth not to show a perceptible parallax. It is evident that for purposes of observation he considered the geometrically determined solar parallax imperceptible. He left the order of Venus and Mercury with respect to each other undecided in the *Almagest*.

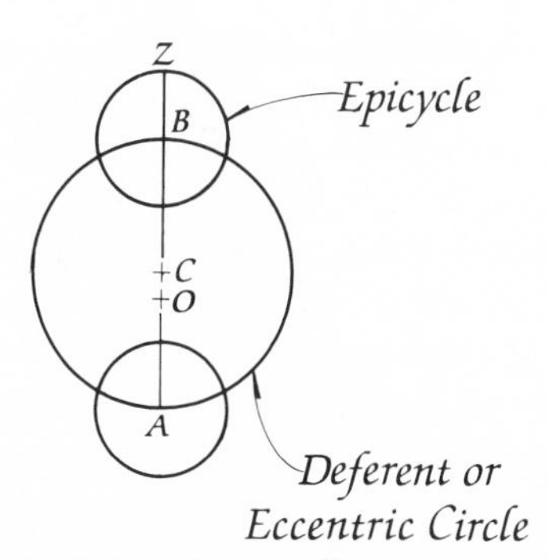

FIG. 7. Ptolemy's model of Venus's motion.
$CB = R = 60$ parts; $BZ = r = 43\frac{1}{6}$ parts; $CO = e = 1\frac{1}{4}$ parts. O is the Earth.

In the *Planetary Hypotheses*, written after the *Almagest*, Ptolemy addressed wider physical and cosmological issues and developed a coherent scheme of sizes and distances of all the heavenly bodies. The *Planetary Hypotheses* consists of two books. The first contains a synopsis of planetary theory and a section on the order, sizes, and distances of the heavenly bodies. The second contains Ptolemy's physical model of epicyclic and eccentric motion. In the Greek version that has come down to us, the second section of book 1, the part containing the scheme of sizes and distances, is missing; it has survived only in Hebrew and Arabic codices. For this reason, the origin of a scheme of sizes and distances that figured prominently in Moslem and Christian medieval cosmology was a mystery to historians.[19] Thanks, however, to the painstaking scholarship of Willy Hartner and Bernard Goldstein, [20] the second part of book 1 was recently restored to its rightful place in the history of cosmology, and the puzzle was solved: the originator of this well-known scheme of sizes and distances was Ptolemy himself.

In discussing the order of the spheres, Ptolemy repeated the statement in the *Almagest* cited above and added that the lack of observed transits of Venus and Mercury across the Sun could be the result of their smallness. Such events would, at any rate, occur very infrequently. The traditional criteria did not allow certain judgment about the order of any of the planets, not even Mars, Jupiter, and Saturn.[21]

At this point, however, Ptolemy introduced an entirely new approach to the problem. From the planetary models in the *Almagest*, one can obtain the ratio of a planet's greatest to least relative distances from the Earth. Thus, in the model for Venus, (see fig. 7), the eccentricity, e, was $1\frac{1}{4}$ parts, and the radius of the epicycle, r, was $43\frac{1}{6}$ parts, while the radius of the deferent or eccentric circle, R, was 60 parts. Venus's apogee distance was $R + r + e$, or $104\frac{5}{12}$ parts, and its perigee distance was $R - r - e$, or $15\frac{7}{12}$ parts.[22] Ptolemy rounded these numbers off and obtained $^{104}/_{16}$ for the ratio of Venus's greatest to least relative distances.

This ratio, he argued, was also the ratio of greatest to least *absolute* distances. As we have seen, he had already done this in the case of the Moon in the *Almagest*, but in the *Planetary Hypotheses* he generalized the physical significance of eccentric and epicyclic constructions which had only mathematical significance in the *Almagest*. To this he now added a proposition from Aristotelian natural philosophy: the cosmos is a plenum, so there can be no empty spaces between the spheres.[23] The greatest geocentric distance of one planet therefore had to equal the least distance of the next higher planet:[24]

> Let us assume that only the spheres of Mercury and Venus lie below the sphere of the Sun, but that the others do not. We have explained in the *Almagest* that the least distance of the Moon is 33 earth radii, and its greatest distance 64 earth radii, dropping fractions. Moreover, the least distance of the Sun is 1,160 earth radii, and its greatest distance is 1,260. The ratio of the least distance of Mercury to its greatest distance is equal to about 34:88, and it is clear from the assumption that the least distance of Mercury is equal to the greatest distance of the Moon, that the greatest distance of Mercury is equal to 166 earth radii, if the least distance of Mercury is 64 earth radii. The ratio of the least distance of Venus to its greatest distance is equal to about 16:104. It is clear from the assumption that the greatest distance of Mercury is equal to the least distance of Venus, that the greatest distance of Venus is 1,079 earth radii, and the least distance of Venus is 166 earth radii.

Obviously, Ptolemy had introduced two further assumptions here. First, he had put Venus above Mercury. The procedure would work the same way if one were to place Mercury above Venus, in which case Mercury's greatest distance would be 1,079 e.r. Second, he had set the unspecified solar distance of 1,210 e.r. found in the *Almagest* equal to the Sun's *mean* distance here, and since the Sun's eccentricity was 2½ parts in 60, its least distance was therefore 1,160 e.r. and its greatest distance 1,260 e.r. The boundary between the spheres of Venus and the Sun presented a problem, however, for Venus's greatest distance arrived at by the nesting sphere procedure did not square with the Sun's least distance determined by means of the eclipse diagram:[25]

> Since the least distance of the Sun is 1,160 earth radii, as we mentioned, there is a discrepancy between the two distances which we cannot account for; but we were led inescapably to the distances which we set down. So much for the two (planetary) spheres which lie closer to the earth than the others. The remaining spheres cannot lie between the spheres of the Moon and the Sun, for even the sphere of Mars, which is the nearest to the earth of the remaining spheres, and whose ratio of greatest to least distance is about 7:1, cannot be accommodated between the greatest distance of Venus and the least distance of

> the Sun. On the other hand, it so happens that when we increase the distance to the Moon, we are forced to decrease the distance to the Sun, and *vice versa*. Thus, if we increase the distance to the Moon [in the eclipse diagram in the *Almagest*] slightly, the distance to the Sun will be somewhat diminished and it will then correspond to the greatest distance of Venus.

Ptolemy did not make this adjustment, but it can easily be done. If, in the above procedure as well as in the determination of the solar distance in the *Almagest* (see eq. [6]), Ptolemy had used a lunar distance of 64.425 e.r., keeping all other parameters the same, both Venus's greatest distance and the Sun's least distance would have come out to be 1,083 e.r. But there was perhaps also another way to make the discrepancy disappear. The ratios of greatest to least planetary distances used in the *Planetary Hypotheses* are rounded off, and the ratio for Mercury is anomalous because from the *Almagest* one obtains the ratio 91.50:33.07.[26] This ratio, a lunar distance of 64⅙ e.r., and the more accurate ratio for Venus found in the *Almagest* yield a greatest distance of 1,189 e.r. for Venus, almost 30 e.r. greater than the Sun's least distance. Obviously, by making minor adjustments of the appropriate kind in the ratios, Ptolemy could have made Venus's greatest distance exactly equal to the Sun's least distance.[27] In the *Almagest* there are no such discrepancies, and the lesser care Ptolemy bestowed on the *Planetary Hypotheses* may very well be a measure of the lesser importance he attributed to it.

The remaining discrepancy must, however, not be allowed to obscure the larger measure of agreement. With only minor differences, the results obtained by three methods—lunar dichotomy, the eclipse diagram, and the nesting sphere procedure—agreed with each other. This confirmation must be seen as a potent argument for accepting the scheme of sizes and distances as well as the methods by which it was constructed.

As the choice of putting Mercury below Venus was still arbitrary, Ptolemy now brought forward an additional reason for his preferred order of the planets: the complexities of their motions. The Moon's motion was the most complex, and Mercury's was next in complexity. The Sun occupied an intermediate position, and the motion of the fixed stars was simplest:[28]

> The spheres nearest to the air move with many kinds of motion and resemble the nature of the element adjacent to them. The sphere nearest to universal motion is the sphere of the fixed stars which moves with a simple motion, resembling the motion of a firm body whose revolution in itself is eternally unchanging.

Having thus disposed of the problem of the planets' order, Ptolemy completed his scheme of celestial distances using the greatest distance of the Sun which followed from the eclipse diagram in the *Almagest*, 1,260 e.r.[29]

> The distances of the three remaining planets may be determined without difficulty from the nesting of the spheres, where the least distance of a sphere is considered equal to the greatest distance of the sphere below it. The ratio of the greatest distance of Mars to its least distance is, again, 7:1. When we set its least distance equal to the greatest distance of the Sun, its greatest distance is 8,820 earth radii and its least distance 1,260 earth radii. The ratio of the least distance of Jupiter to its greatest distance is equal to the ratio 23:37. When we set the least distance of Jupiter equal to the greatest distance of Mars, its greatest distance is 14,187 [read 14,189] earth radii and its least distance 8,820 earth radii. Similarly, we set the ratio of the least distance of Saturn to its greatest distance equal to the ratio 5:7, and the least distance of Saturn equal to the greatest distance of Jupiter. Therefore the greatest distance of Saturn, which is adjacent to the sphere of the fixed stars, is 19,865 earth radii, and its least distance is 14,187 [read 14,189] earth radii.
>
> In short, taking the radius of the spherical surface of the earth and the water as the unit, the radius of the spherical surface which surrounds the air and the fire is 33, the radius of the lunar sphere is 64, the radius of Mercury's sphere is 166, the radius of Venus' sphere is 1,079, the radius of the solar sphere is 1,260, the radius of Mars' sphere is 8,820, the radius of Jupiter's sphere is 14,18[9], and the radius of Saturn's sphere is 19,865.

Ptolemy's value for the radius of the Earth was "two myriad stades and a half and a third and one part in thirty myriad stades," that is, 28,667 stades.[30] He then converted the above distances to stades, concluding with the distance to the fixed stars:[31]

> The boundary that separates the sphere of Saturn from the sphere of the fixed stars lies at a distance of 5 myriad myriad and 6,946 myriad stades and a third of a myriad stades.

In other words, the distance to the fixed stars, the radius of the cosmos, was 569,463,333 stades, let us say roughly 50 million modern miles. Ptolemy had now expressed all the distances in the heavens, as well as the extent of the cosmos itself, in terms of units used in daily life to measure distances.

His next step was to determine the apparent sizes of the heavenly bodies so he would be able to calculate their absolute sizes. Here again he drew to some extent on the work of Hipparchus:[32]

> Hipparchus said that the apparent diameter of the Sun is 30 times as great as that of the smallest star, and that the apparent diameter of Venus, which appears to be the largest star, is about a tenth the apparent diameter of the Sun. . . . Hipparchus did not make clear at which distance of Venus its diameter takes on

> the value quoted, but we consider this amount to be its apparent diameter at mean distance where the planet is usually seen, for at apogee and perigee it is hidden by the rays of the Sun. We too find that the apparent diameter of Venus is a tenth that of the Sun, as Hipparchus stated. Moreover, we find the diameter of Jupiter to be 1/12 the diameter of the Sun; Mercury's 1/15 the diameter of the Sun; Saturn's 1/18 the diameter of the Sun; and the diameter of Mars, and of first magnitude stars, 1/20 the diameter of the Sun. The diameter of the Moon at mean distance on its sphere, and mean distance of the eccentric sphere, is equal to 1 1/3 times the diameter of the Sun.

In giving all apparent diameters at mean distances, Ptolemy arrived at a very large apparent diameter of the Moon. If the Moon's apparent diameter is equal to that of the Sun at the Moon's apogee distance (64 1/6 e.r.), then at mean distance (48 e.r.) it should be $64/48 = 1\,1/3$ times the Sun's apparent diameter. This very large and erroneous angular diameter given in the *Planetary Hypotheses* demonstrates that Ptolemy accepted the physical consequences of his mathematical lunar theory in the *Almagest*.

Ptolemy gave no indication of how he had measured the apparent diameters of the planets and fixed stars. A direct comparison with the Sun is impractical; direct comparison with the disk of the Moon seems more reasonable. It was, at any rate, best to express the comparison in terms of the Sun's apparent diameter, since no variation had been detected in it.

Knowing the apparent diameters of all heavenly bodies, as well as their absolute distances, Ptolemy could now calculate their actual diameters and volumes. In the *Almagest* he had determined the Sun's actual diameter to be 5 1/2 times the Earth's.[33] He now multiplied this number by the ratio of the planet's mean distance to the Sun's mean distance, and by the ratio of the planet's apparent diameter to the Sun's apparent diameter. Thus, Jupiter's actual diameter was $5\,1/2 \times 11{,}503/1{,}210 \times 1/12 = 4.36$ times the Earth's, or, as Ptolemy put it, $4\,1/3 + 1/40$. The ratios of volumes were the cubes of the ratios of the diameters.[34]

> . . . we find that in the measure where the diameter of the earth is 1, the diameter of the Moon is 1/4 [+] 1/24; the diameter of Mercury 1/27; the diameter of Venus 1/4 + 1/20; the diameter of the Sun 5 1/2, the diameter of Mars 1 1/7; the diameter of Jupiter 4 1/3 + 1/40; the diameter of Saturn 4 1/4 + 1/20; and the diameters of the fixed stars of the first magnitude at least 4 1/2 + 1/20. In the measure where the volume of the earth is 1, the volume of the Moon is 1/40; the volume of Mercury 1/19,683; the volume of Venus 1/44; the volume of the Sun 166 1/3; the volume of Mars 1 1/2; the volume of Jupiter 82 1/2 + 1/4 + 1/20; the volume of Saturn 79 1/2; the volume of first magnitude stars at least 94 1/6 + 1/8. Accordingly the Sun has the greatest volume, followed by the fixed stars of the first magnitude. The third in rank is Jupiter, the fourth Saturn, the fifth Mars, the sixth earth, the seventh Venus, the eighth the Moon, and lastly Mercury.

In conclusion, Ptolemy mentioned the implications of this scheme for the parallax of the planets, which he had considered imperceptible in his discussion in the *Almagest*. Obviously, if Mercury's sphere was directly above the sphere of the Moon, Mercury should have a sizable parallax at its perigee, equal to the Moon's apogee distance. Ptolemy wrote:[35]

> We now repeat that, if all the distances have been given correctly, the volumes are also in accord with what we have said. If the distances are greater than those we described, then these sizes are the minimum values possible. If their distances are correctly given, Mercury, Venus, and Mars display some parallax. The parallax of Mars, at perigee, is equal to that of the Sun at apogee. The parallax of Venus at apogee is close to that of the Sun at perigee. The parallax of Mercury at perigee is equal to that of the Moon at apogee, while the parallax of Mercury at apogee is equal to that of Venus at perigee. The ratio of each of them to the lunar and solar parallax is equal to the ratio of the distances that we have mentioned to the distances of the Sun and the Moon.

It is interesting to note that Ptolemy hinted here that perhaps he had determined the *minimum* distances of the planets. We may infer that he meant that although there might be spaces between the spheres, such as the one he left between Venus and the Sun, the smallest possible cosmos that followed from the *Almagest* was one without such spaces. According to the present scheme, Mercury should show a perceptible parallax, but this planet is difficult to observe because of its proximity to the Sun. At Venus's closest approach to the Earth, when its parallax might be detectable with the instruments available to Ptolemy, it is in conjunction with the Sun. This planet, like Mercury, is best observed when it is at its greatest elongation from the Sun, and there it is nearly at its mean distance from the Earth (622 e.r.). Its parallax should be about twice the Sun's, or 6′, a quantity not detectable in this kind of measurement. There was therefore no convenient way to follow up Ptolemy's suggestion. Virtually all his successors until the sixteenth century ignored it, taking the scheme of sizes and distances in table 1 not as the minimum but the actual measure of the cosmos.

The Ptolemaic System of cosmic dimensions was an ingenious mélange of philosophical tenets, geometric demonstrations with spuriously accurate parameters, planetary theories, and naked-eye estimates. It was a speculative by-product of the first complete system of mathematical astronomy. Although in retrospect the distances were in error by an order of magnitude, the values found by the nesting sphere principle were confirmed by the distance of the Sun found in the *Almagest* and also by Aristarchus's ratio of solar to lunar distance. The naked-eye estimates of the apparent diameters of fixed stars and planets seemed reasonable enough. The resulting scheme of sizes and distances was, therefore, as plausible as

TABLE 1 The Ptolemaic System of Sizes and Distances

Body	Ratio of Perigee to Apogee Distances	Absolute Distance in e.r.			Apparent Diameter at Mean Distance (Sun = 1)	Actual Diameter (Earth = 1)	Volume (Earth = 1)
		Least	Greatest	Mean			
Moon	33:64	33	64	48	1⅓	¼ + 1/24	1/40
Mercury	34:88	64	166	115	1/15	1/27	1/19,683
Venus	16:104	166	1,079	622½	1/10	¼ + 1/20	1/44
Sun[a]	[57½:62½]	1,160	1,260	1,210	1	5½	166⅓
Mars	1:7	1,260	8,820	5,040	1/20	1⅐	1½
Jupiter	23:37	8,820	14,187[b]	11,503	1/12	4⅓ + 1/40	82½ + ¼ + 1/20
Saturn	5:7	14,187[b]	19,865	17,026	1/18	4¼ + 1/20	79½
Fixed stars	—	—	—	20,000[c]	1/20	4½ + 1/20	94⅙ + ⅛

[a]The Sun's eccentricity is given in the *Almagest* as 2½ parts in 60.

[b]This is an error: 37/23 × 8,820 = 14,189 = 5/7 × 19,865. Since 14,189 was obviously used to calculate Saturn's greatest distance, 14,187 must be a scribal error.

[c]In computing the actual diameter of a first magnitude star, Ptolemy used a distance of 20,000 e.r.

it was ingenious, and it came to occupy an important place in cosmological thought among Moslem astronomers and then the Christian schoolmen.[36]

Among Ptolemy's immediate successors, however, this system seems not to have been very important. Astronomers were more interested in astrological predictions than in physical explanations;[37] philosophers ignored quantitative matters. Indeed, Ptolemy's authorship of the *Planetary Hypotheses* was apparently unknown among some commentators within a few centuries after his death. The nesting sphere principle was used by philosophers in their discussions of the order of the planets, however. Thus, in his *In Platonis Timaeus Commentaria*, Proclus (A.D. 410–85) gave the distances of the Moon, Mercury, Venus, and the Sun found in the *Planetary Hypotheses*, but in his *Hypotyposis* he used distances computed from the ratios found in the *Almagest*, ascribing the principle of nesting spheres to "certain authorities."[38] Simplicius (6th century A.D.) gave no numbers but ascribed the nesting sphere principle to the *Almagest*![39]

The *Planetary Hypotheses* and the Ptolemaic system contained in it were, then, not very important in the astronomy and cosmology of late antiquity. The all-important second part of book 1 was probably already missing from many editions at an early date, and this would explain why it did not survive in the Greek version.[40] However, when Moslem astronomers began to build on Ptolemy's achievement seven centuries after his death, they gave the Ptolemaic System a prominent place in their works. At that time the important career of the system began.

4
The Ptolemaic System Enshrined

A neophyte wishing to enter the world of European medieval literature faces daunting obstacles, of which the language barrier is by no means the most serious one. The intellectual world of medieval man is now so foreign to us that the beginning student needs a guide to help him get his bearings. One of the most celebrated guides of the last several generations was C. S. Lewis (1898–1963), whose posthumous book *The Discarded Image: An Introduction to Medieval and Renaissance Literature* (1964) became a classic overnight.

In *The Discarded Image* Lewis described for the beginner "the imagined universe which is usually presupposed in medieval literature and art."[1] He characterized medieval man as "an organiser, codifier, a builder of systems," who wanted "a place for everything and everything in the right place."[2] In his quest to harmonize all the apparent contradictions among the various disparate parts of his intellectual heritage, medieval man constructed a "single, complex, harmonious mental Model of the Universe."[3] Lewis claimed that knowledge of this model is fundamental to an understanding of medieval literature:[4]

> I hope to persuade the reader not only that this Model of the Universe is a supreme medieval work of art but that it is in a sense the central work, that in which most particular works were imbedded, to which they constantly referred, from which they drew a great deal of strength.

When it came to the dimensions of this model, however, Lewis was curiously uninformed:[5]

> The dimensions of the medieval universe are not, even now, so generally realised as its structure; . . . The reader of this book will already know that Earth was, by cosmic standards, a point—it had no appreciable magnitude. The stars, as [Cicero's] *Somnium Scipionis* had taught, were larger than it. Isidore in the sixth century knows that the Sun is larger, and the Moon smaller than the Earth (*Etymologies*, III, xlvii–xlviii), Maimonides in the twelfth maintains that every star is ninety times as big, Roger Bacon in the thirteenth simply that the least star is 'bigger' than she.

Had Lewis actually looked at Roger Bacon's *Opus Maius* instead of using a secondary source, he would have found the Ptolemaic System in all its detail laid out before him.[6] And Bacon's rendition of it was by no means an isolated incidence: the Ptolemaic cosmic dimensions can be found throughout the spectrum of the literature of the High Middle Ages, from the technical to the popular.

Lewis's main sources for *The Discarded Image* date from the first century B.C. to the sixth century A.D., from Cicero's *Dream of Scipio* (a part of his *Republic*) to Boethius's *Consolation of Philosophy*.[7] Unfortunately, the Ptolemaic System appears in none of these works; Ptolemy's complete scheme of cosmic dimensions entered Europe more than half a millennium after Boethius, through the astronomical literature of the Moslems.

* * *

Astronomy did not flourish after Ptolemy. As Christianity came to dominate the spiritual life in the Mediterranean world, interest in esoteric pagan learning was increasingly relegated to the peripheral areas, such as Syria and Persia. The first revival of mathematical astronomy occurred after the rise of Islam among scholars of various ethnic and linguistic origins, who were attracted to the Abbasid court in Baghdad beginning in the eighth century A.D. In the ninth century at least three translations of the *Almagest* into Arabic were made; the most influential one was made by the translator Ishaq ibn Hunain (d. 910–11) and revised by the astronomer Thābit ibn Qurra (827–901).[8] Ptolemy's *Planetary Hypotheses* was also rendered into Arabic in the same century under the title *Kitāb al-Manshūrāt*, and evidence suggests that Thābit was responsible for this translation.[9] Although the Ptolemaic System contained in *Planetary Hypotheses* became part of the common stock of knowledge among Moslem astronomers, the fact that many of them were not aware of the system's origin indicates that the book itself did not gain great currency in the scholarly community of Islam.

The *Almagest* is a difficult book, not suited as a text for introducing beginners to Ptolemaic astronomy. For this reason a number of elementary instructional astronomy books were produced in Islam and later in the Latin West. One of the earliest textbooks for beginners was written by the astronomer-astrologer al-Farghānī (ca. 800–870). His "summary" or *Jawāmiᶜ* of the *Almagest* was a very simple, qualitative introduction to Ptolemaic astronomy; it achieved great popularity.

Although al-Farghānī was familiar with the nesting sphere principle and knew the figures for the apparent diameters of the planets, he did not know the *Planetary Hypotheses* and did not ascribe the Ptolemaic System to Ptolemy. He computed the absolute distances directly from the *Almagest* and arrived at the results shown in table 2.[10]

In computing the ratio of Mercury's greatest to least distance from the model in the *Almagest*, al-Farghānī made an error which has been reconstructed by Swerdlow. The complicated model of Mercury has the curious property of producing two perigees, each about 120° removed from the apogee.[11] The correct ratio derived from this model is 91.50:33.07, but al-Farghānī computed the relative distance of the planet when it is 180°

TABLE 2 Al-Farghani's Scheme of Sizes and Distances

Body	Absolute Distance in e.r.[a]			Apparent Diameter at Mean Distance	Actual Diameter	Volume
	Least	Greatest	Mean	(Sun = 1)	(Earth = 1)	(Earth = 1)
Moon	32½ + 1/20	64⅙	48⅚	[1⅓]	1/3⅖	1/39
Mercury	64⅙	167	115½	1/15	1/28	1/22,000
Venus	167	1,120	643½	1/10	1/3⅓	1/37
Sun	1,120	1,220	1,170	1	5½	166
Mars	1,220	8,876	5,048	1/20	1⅙	1½ + ⅛
Jupiter	8,876	14,405	11,640	1/12	4½ + 1/16	95
Saturn	14,405	20,110	17,258	1/18	4½	91
Fixed stars						
M1	—	—	20,110	1/20	4½ + ¼	107
M2	—	—	20,110	—	—	90
M3	—	—	20,110	—	—	72
M4	—	—	20,110	—	—	54
M5	—	—	20,110	—	—	36
M6	—	—	20,110	—	—	18

[a]1 e.r. = 3,250 miles. Al-Farghani gives all distances in miles as well.

from the apogee (which for any other planet would be the perigee) and produced the ratio 91.50:34.50, which he rounded off to 13:5. Thus, instead of obtaining 177 e.r. as Proclus had in his *Hypotyposis*, al-Farghānī obtained 167 e.r. for Mercury's greatest distance.[12] The fact that this value was very close to Ptolemy's 166 e.r. found in the *Planetary Hypotheses* is entirely fortuitous and does not mean that al-Farghānī was familiar with this work.

This procedure led him to a greatest distance of Venus of 1,120 e.r., and at this point al-Farghānī did not introduce the solar distance from the *Almagest*. Instead, he made 1,120 e.r. the least solar distance and arrived at a greatest solar distance of 1,220 e.r.,[13] which is, of course, very close to the unspecified solar distance found by Ptolemy by means of the eclipse diagram in the *Almagest*. Al-Farghānī's scheme of distances thus had no gaps or overlaps in it. His cosmos was slightly larger than Ptolemy's, and he added the refinement of computing the volumes of the fixed stars down to the sixth magnitude by dividing the volume of a first magnitude star (107 earth volumes) by six (see table 2).

Under al-Ma'mun astronomers had measured the length of a degree, and their result was 1° = 56⅔ miles; the Earth's radius was therefore 3,250 miles. Al-Farghānī used this measure to convert all his distances into miles. The mile used here consisted of 4,000 "black cubits" of 24 fingerbreadths each.[14] Al-Farghānī's scheme was to become very influential in the West.

Another elementary introduction to the *Almagest*, *Tashil al-Majisti*, is attributed to Thābit ibn Qurra. This tract contains the measures found in the *Planetary Hypotheses* (its translation into Arabic, as mentioned earlier, may also be the work of Thābit) with two exceptions. Thābit, if he was the author, eliminated the gap between Venus and the Sun by simply letting the Sun's least distance equal Venus's greatest distance, 1,079 e.r., but he kept Ptolemy's value for the Sun's greatest distance, 1,260 e.r. The Sun's implied mean distance therefore would be 1,169½ e.r., and its eccentricity would be about 4⅔ part in 60, a value much larger than Ptolemy's. Thābit's solution to the problem was thus not very satisfactory. All other distances are the same as Ptolemy's, as are the volumes except for Venus's, which Thābit made 1/37th instead of 1/44th of the Earth's. This may be a later intrusion of al-Farghānī's value.[15]

If the efforts of al-Farghānī and Thābit ibn Qurra were simply teaching aids, symptomatic of the difficulty of the *Almagest* for beginners, Moslem scholars quickly began to produce works that were based on research as well. Al-Battānī (ca. 850–929), the son of an instrument maker, made many observations and produced a knowledgeable commentary on and supplement to the *Almagest*, his *Zīj*. Although al-Battānī's thoughts on planetary sizes and distances were based on the *Planetary Hypotheses*, he ascribed the nesting sphere principle, as well as the distances based on it, to "modern eminent philosophers . . . after Ptolemy."[16] However, he did not hesitate to make changes in the sizes and distances.

Al-Battānī attempted to eliminate the gap between Venus and the Sun by recalculating the solar distance by means of the eclipse diagram. He did so by changing the lunar distance at which the Moon's apparent diameter is equal to the Sun's from Ptolemy's 64⅙ e.r. to 60 29/30 e.r., while keeping all other parameters the same as Ptolemy's. Equation (6) on page 18 thus becomes:

$$ND = \frac{60\tfrac{29}{30}}{3\tfrac{3}{5} \times 60\tfrac{29}{30} \times \sin 15'40'' - 1}$$

These values would produce a solar distance of about 250,000 e.r., an absurdly large measure that would clearly rule out the nesting sphere arrangement. Al-Battānī did not show the calculation; instead he multiplied the lunar distance of 60 29/30 e.r. by the ratio of Ptolemy's solar to lunar distance in the *Almagest*, 1,210:64⅙, rounded off to 18⅘:1. The resulting solar distance was 1,146 e.r. (corresponding to a parallax of exactly 3′!), which al-Battānī set equal to the Sun's greatest distance. Having found the Sun's eccentricity to be 2 1/12 part in 60, he made the Sun's least distance 1,070 e.r. and its mean distance 1,108 e.r. (see table 3).[17]

TABLE 3 Al-Battānī's Scheme of Sizes and Distances

Body	Absolute Distance in e.r.			Apparent Diameter at Mean Distance	Actual Diameter	Volume
	Least	Greatest	Mean	(Sun = 1)	(Earth = 1)	(Earth = 1)
Moon	33½ + 1/20	64⅙	48⅚	[1⅓]	1/3⅖	1/39¼
	64⅙	166	115½	1/15	1/26¼	1/18,087
Mercury	166	1,070	618	1/10	¼ + 1/20	1/36
	1,070	1,146	1,108	1	5½	166⅜
Venus	1,146	8,022	4,584	1/20	1 1/7	1⅓
Sun	8,022	12,924	10,473	1/12	4⅓	81
Mars	12,924	18,094	15,509	1/18	4⅙ + ⅛	79
Jupiter						
Saturn						
Fixed	—	—	19,000	1/20	4⅔ + 1/20	105
stars	—	—	19,000	—	—	16
M1						
M6						

Al-Battānī's scheme of sizes and distances was based on the same ratios of greatest to least distances as those found in the *Planetary Hypotheses*, except for the Sun's. He should have found Venus's greatest distance to be 1,079 e.r., but he set it equal to the Sun's least distance of 1,070 e.r. without further comment. The distances of the superior planets were based on his greatest solar distance of 1,146 e.r., so they were somewhat smaller than those found in the *Planetary Hypotheses*. Al-Battānī's apparent diameters were the same as Ptolemy's, and his actual diameters and volumes differed only slightly from Ptolemy's.[18]

Other Moslem astronomers, including the great al-Bīrūnī (973–1048),[19] incorporated the Ptolemaic System in their works. Although almost all of them made changes in the distances (but not in the apparent diameters), these changes were always minor. Within the accepted framework of the nesting spheres, which was questioned by no one, they tried to eliminate imperfections, especially the gap between Venus and the Sun.

There was at least one exception to this consensus on sizes and distances, however. Some time before he became associated with the new Maraghah Observatory in 1259, the astronomer al-ᶜUrḍī (d. 1266) wrote a treatise, *Kitāb al-Hay'ah*, in which he developed non-Ptolemaic planetary models. Only two known copies of this important tract, which marks the beginning of the so-called Maraghah school of astronomy, have been found so far.[20] In *Kitāb al-Hay'ah*, al-ᶜUrḍī disagreed firmly with Ptolemy's procedure in the *Planetary Hypotheses*.

Ptolemy had taken all the apparent planetary diameters from Hipparchus, according to al-ᶜUrḍī, and since Hipparchus had not said at what distances he had measured them, Ptolemy had assumed that they were measured at mean distances. This created a problem, however. If, for instance, Venus's apparent diameter were $\frac{1}{10}$th the Sun's at Venus's mean distance of $622\frac{1}{2}$ e.r., it should be $622.5/166 \times \frac{1}{10}$ or (about) $\frac{2}{5}$th the Sun's apparent diameter at its perigee distance. This is clearly not the case, and al-ʿUrḍī therefore concluded that Hipparchus had measured the planetary diameters at their *least* distances.[21]

Al-ʿUrḍī recalculated all the distances from parameters which he redetermined, making sure to add the diameter of each heavenly body to the calculated thickness of its sphere. He found that there was not enough room between Mercury and the Sun for the sphere of Venus. Venus therefore had to be above the Sun, so the order of the planets became Moon-Mercury-Sun-Venus-Mars-Jupiter-Saturn. Whereas Ptolemy had used distances that showed that Mercury and Venus had to be below the Sun, al-ʿUrḍī had used the same argument to show that Venus had to be above the Sun. Since his greatest solar distance was 1,266 e.r., about the same as Ptolemy's, Venus's greastest distance became 8,486 e.r., which was Mars's least distance. The spheres of Mars, Jupiter, and Saturn were thus expanded enormously, and the radius of the cosmos turned out to be 140,147 e.r., that is, more than 7 times as great as Ptolemy's! What is more, all the diameters except the Sun's changed dramatically as well. For example, instead of being $4\frac{1}{4} + \frac{1}{20}$ times as great as the Earth's, Saturn's actual diameter was now $25\frac{13}{30}$ times the Earth's.[22]

As bold as his reworking of the numbers was, al-ʿUrḍī had not broken with the principle of the nesting spheres, and he had also not questioned Hipparchus's and Ptolemy's "measurements" of the apparent planetary diameters. His sizes and distances have thus far not been found in any other Moslem or Latin source; only the affectation of adding the planetary diameters to the thicknesses of their spheres found an echo elsewhere, as we shall see later.

* * *

When, with the revival of learning, mathematical astronomy entered the West, Latin scholars drew heavily on Moslem textbooks to help them understand Ptolemaic astronomy. John of Seville translated al-Farghānī's *Jawāmiʿ* into Latin in 1137, giving it the title *Differentie Scientie Astrorum*, while Gerard of Cremona rendered it into Latin under the title *Liber de Aggregationibus Scientie Stellarum et Principiis Celestium Motarum* in 1175, the same year he translated the *Almagest*. Al-Battānī's more advanced *Zīj* had already been translated by Plato of Tivoli earlier in the century under the title *De Motu Stellarum*.[23] Moslem textbooks and commentaries thus preceded the *Almagest*, and we may assume that the vast majority of Latin scholars learned Ptolemaic astronomy from elementary textbooks and not from the *Almagest* itself, which was studied by experts only.

Latin scholars were thus exposed to the Ptolemaic System of sizes and distances from the very beginning, and this scheme quickly became a part of the common stock of knowledge of literate people in the West. Since al-Farghānī's textbook was by far the most popular of these early astronomical primers and a model for later textbooks written by Latin schoolmen, it was al-Farghānī's particular version of the Ptolemaic System that became the accepted or canonical one in the West. Often the fate of an idea so common that it is usually taken for granted by writers of the period is that its importance may escape readers of a later period. In the history of science this scheme of sizes and distances, so important in medieval and early modern cosmology, has to a large degree met this fate. We find it discussed today only in works that deal specifically with the astronomy of that period, with very few exceptions.[24] Yet, if we examine medieval literature, we find the Ptolemaic System almost everywhere.

In medieval astronomical literature the scheme is, of course, found in the many copies of al-Farghānī's work, *Differentie* or *Liber de Aggregationibus*.[25] In the thirteenth century this work was replaced by indigenous textbooks of astronomy, of which *De Sphaera* by Johannes Sacrobosco became the most popular. This little book, however, was so elementary that the dimensions of the heavens were beyond its scope. Sizes and distances are found in commentaries on the *Sphere* of Sacrobosco and in more advanced treatises, such as Campanus of Novara's *Theorica Planetarum* composed around 1260.

Campanus accepted the principle of the nesting spheres, and he gave the elements of each planetary model (deferent radius, epicycle radius, etc.) in miles. Using al-Farghānī's parameters, he recalculated the planetary distances in miles, adding (as did al-ʿUrḍī) the planetary diameters to the thicknesses of the spheres. Campanus also used the more precise value of 3,245 5/11 miles for the Earth's radius.[26] Consequently, he arrived at distances that were slightly different from those of al-Farghānī. Saturn's greatest distance, for instance, came out to be 73,387,747 miles, that is, 22,612 e.r., instead of al-Farghānī's 20,110 e.r.[27] Campanus's scheme was followed, without explanation or acknowledgment, by Robertus Anglicus in his commentary on Sacrobosco's *Sphere*, composed in 1271.[28] The numerous surviving copies of the works by Campanus and Robertus indicate their importance in medieval education.[29]

In the *Compilatio de Astrorum Scientia*, composed at about the same time, the otherwise unknown author, Leopold of Austria, gives the sizes and distances with some variations. The work was translated into French before 1324, and as one of the earliest astronomy textbooks in the vernacular it enjoyed great popularity over the next several centuries. Leopold wrote:[30]

> Les grandeces des cors celestres sont 14. Li Solaus a le quantitet de le Terre 100 fois 60 fois 6 fois et le 4ᵗ d'une 8ᵉ, les plus grandes estoilles en ont 115 fois, Jupiter 95 fois, Saturne 91

> fois, chelles qui sont de le seconde quantité fixes estoilles 90 fois, chelles qui sont de le tierche quantitet 70 fois, de le quarte 50 fois, de le quinte 36 fois, de le sixte 18 fois, Mars une fois et le moitiet d'une 8ᵉ par une fois, Venus au rewart de le Terre le porcion de 37, li Lune 39 et un poi plus, Mercure une part de 22,000 partie de le Terre.
>
> Les longaices des estoilles de Terre: les longaiches des cors celestres de Terre sont 8. Li dyametres de le Terre a 6500 milles. . . . De le Terre jusques au premier lieu de le Lune sont les moitiés du dyametre de le Terre 33, dusques a Mercure 64, dusques a Venus 166, dusques au Soleil 1079, dusques a Mars 1260, dusques a Jupiter 8820, dusques a Saturne 1[4], 187, dusques a fixes estoilles a 1[9],[8]6[5].

The sizes are clearly derived from al-Farghānī (see table 2) and can be found in Thābit's *De Quantitatibus*, while the distances are clearly those of Ptolemy (see table 1) and can be found in Thābit's *De Hiis*.[31]

The cosmic dimensions are also found in the more general philosophical literature of the period. Among a number of instances,[32] the most celebrated rendition of al-Farghānī's scheme is the version found in the *Opus Maius* of Roger Bacon (1214–92) composed in 1266. After giving the dimensions of the Earth, Bacon writes:[33]

> After looking into these matters we must consider the altitude of the heavenly bodies, and likewise their size and thickness. For Alfraganus says in the twenty-first chapter that Ptolemy and other scientists took the semidiameter [of the Earth] as the length with which they measured distances from the center of the earth, and took the magnitude of the earth as the quantity with which they measured the magnitudes of the stars. This is quite clear from the demonstrations of Ptolemy in the fifth section of the *Almagest*. Therefore Alfraganus is of the opinion, from a comparison of the semidiameter of the earth with that of the starry sphere, that the distance of the starry sphere from the center of the earth is 20,110 times the semidiameter of the earth, 65,357,500 miles, which if doubled will give the diameter of the whole starry sphere as 130,715,000 miles. When this is multiplied into three and one seventh we shall have the circumference of a great circle in the starry sphere, namely, 410,818,517 miles, and three sevenths of a mile, that is, 1714 cubits, and two sevenths of a cubit.

With what Arthur O. Lovejoy called "unwearying enthusiasm," Bacon continues to "dilate"[34] on the magnitude of the sphere of the fixed stars, giving the reader the length of a degree on this circle, as well as the surface area of the sphere. Thereupon he proceeds to give the distances of the planets:[35]

> Moreover, the semidiameter of the starry heaven is the longer distance of the sphere of Saturn, because they join without in-

termediate space. But its nearer distance to the earth is 46,816,250 miles, which is the longer distance of the sphere of Jupiter, whose nearer distance is 28,847,000 miles, which is the longer distance of the sphere of Mars, whose nearer distance is 3,965,000 miles, which is the longer distance of the sphere of the sun, whose nearer distance is 3,640,000, which is the longer distance of the sphere of Venus, whose nearer distance is 542,750, which is the longer distance of Mercury, whose nearer distance is 208,541 and two thirds of a mile, and this is 2666 cubits and two thirds of a cubit, and this is the longer distance of the moon, and this, as Alfraganus says, is sixty-four times and a sixth of one time equal to the half of the diameter of the earth, and the nearer distance of the moon is 109,037 and a half of a mile, that is, 2000 cubits, and this distance is thirty-three times and a half of a tenth, that is, one twentieth of one time equal to half of the diameter of the earth. The diameters of the separate spheres are obtained by doubling the semidiameter; the circumference of each is found by tripling the diameter with the addition of a seventh part, and the whole surface of each sphere is found by multiplying its diameter into its circumference, as was explained in the case of the earth and of the starry sphere. Any one can find these dimensions by computation, and for this reason I omit them to avoid prolixity.

Bacon next gives the thicknesses of the spheres in miles and calculated how long it would take a man walking 20 miles per day to reach the sphere of the Moon.[36] A few pages later, he gives the diameters and volumes of the planets and fixed stars according to al-Farghānī, ending with the hierarchy of heavenly bodies according to size:[37]

From all these facts, then, that have been mentioned in regard to the magnitudes of the heavenly bodies it is evident that greater than all, with the exception of all spheres other than that of the earth, is the sun: then in the second place are stars of the first magnitude; in the third place, Jupiter; fourth, Saturn; fifth, all the remaining fixed stars according to their grades and orders; sixth, Mars; seventh, fixed stars known to sight; eighth, the earth; ninth, Venus; tenth, the moon; eleventh, Mercury.

By "fixed stars known to sight" Bacon meant "other stars in infinite number, the size of which cannot be ascertained by instruments, and yet they are known by sight, and therefore have sensible size with respect to the heavens." The context here, as well as in the works of others, was the demonstration that "the Earth does not possess any sensible size with respect to the heavens."[38]

In view of the broad scope and the popularity of Bacon's *Opus Maius*, the widespread knowledge of al-Farghānī's work, and the popularity of textbooks such as Campanus's *Theorica Planetarum*, Robertus's commentary on Sacrobosco's *Sphere*, and Leopold's *Compilatio* and its translation into the vernacular, it is fair to say that virtually all educated persons after about 1250 were familiar with the principle of nesting spheres and the cosmic dimensions derived from it. But if the Ptolemaic System became a commonplace, a part of a deep-seated conception of the cosmos, we might expect to find traces of it in the popular literature of the time as well. This is indeed the case.

As early as 1245, in a French vernacular poem entitled *Image du Monde*, a certain Gossouin of Metz made liberal use of al-Farghānī's cosmic dimensions. The work, in a slightly later prose version, attained immense popularity which continued unabated well into the sixteenth century. It was translated into English and printed by William Caxton as early as 1480, under the title *Mirrour of the World*, and went through numerous editions. A popular encyclopedia, *Mirrour of the World* contained a liberal measure of astronomical information designed to drive home the moral lesson of the vastness of God's creation. In the chapter on "How the Mone and the Sonne haue eche of them their propre heyght," we find the following information:[39]

> The Earth is "more grete than the body of the mone . . . by xxix [read xxxix] tymes and a lytil more." The Moon is "in heyght aboue the erthe xxiiii tymes and an half as moche as therthe hath of thycknes." [That is, the Moon's mean distance is 49 e.r.]
>
> The Sun is "gretter than al therthe wythout ony defaulte by an C.lxvi. tymes, and thre partyes of the xx parte of therthe" [i.e., $166\frac{3}{20}$, by volume].
>
> The mean distance of the Sun is "ffyue hondred lxxx and v tymes as moche as therthe may haue of gretenes and thycknes thurgh" [i.e., 1,170 e.r.].

In the next chapter Gossouin gave the distance of the fixed stars—al-Farghānī's 20,110 e.r.—and brought the measure down to the experience of everyday life:[40]

> Fro therthe vnto the heuen, wherin the sterres ben sette, is as moche grete espace; ffor it is ten thousand and .lv. sythes as moche, and more, as is alle therthe of thycknes [10,055 e.d. = 20,110 e.r.]. And who that coude acompte after the nombre and fourme, he myght knowe how many ynches it is of the honde of a man, and how many feet, how many myles, and how many Journeyes it is from hens to the firmament or heuen. Ffor it is as moche way vnto the heuen as yf a man myght goo the right way without lettyng, and that he myght goo euery day

> xxv myles of Fraunce, . . . and that he taried not on the waye, yet shold he goo the tyme of seuen .M.i.C. and .lvii. yere and a half [i.e., 7,157½ years] er he had goon somoche way as fro hens vnto the heuen where the sterres be inne.

In fact, had Adam started on such a journey the day he was created, he would still have had 713 years to go when this tract was written in 1245.[41] Gossouin had given the length of the year as 365¼ days and the diameter of the Earth as 6,500 miles.[42] The calculation of the length of Adam's journey is correct to within the accuracy of computation, and the date of creation implied is 5199½ B.C., the accepted date for that event.[43]

Even in popular literature as far removed from astronomy as the legends of the saints, we find traces of the cosmic dimensions. In *The South English Legendary*, composed in the vernacular in the thirteenth century, we learn in the legend of St. Michael that the Earth is smaller than the smallest star,[44] and then, [45]

> Biside þe eorþe in þe on half. þe sonne sent out hure liȝt
> An hondred siþe & fiue and sixti. as it is iwrite
> þe sonne is more þanne þe eorþe. woso wolde iwite
> And þe eorþe is more þanne þe mone. nye siþe iwis
> þe mone þinnþ þe more. for he so nei us is
> þe sonne is herre þan þe mone. more þanne suche þreo
> þenne it beo henne to þe mone. þe lasse he is to seo
> Muche is bitwene heuene & eorþe. for þe man þat miȝte go
> Euerich dai forti mile. euene upriȝt and eke mo
> He ne ssolde to þe heiost heuene. þat þe aldai iseoþ
> Come in eiȝte þousond ȝer. þere as þe sterren beoþ
> And þei Adam oure ferste fader. hadde bigonne anon
> þo he was ferst imad. toward heuene to gon
> And hadde euerich day forti mile. euene upriȝt igo
> He hadde noȝt ȝute to heuene icome. bi a þousond ȝer & mo.

The writer should have used day trips of 20 miles, not 40 miles, but the confusion is understandable since the length of the mile and league varied enormously from place to place. Al-Farghānī had used a mile of 4,000 cubits,[46] but in the same demonstration Moses Maimonides (1135–1204) used a mile half as long as al-Farghānī's:[47]

> Accordingly, it has been demonstrated that the distance between the center of the Earth and the highest part of the sphere of Saturn is one that could be covered in approximately eight thousand and seven hundred years of three hundred and sixty-five days each, if each day a distance is covered of forty of our legal miles, of which each has two thousand of the cubits used for working purposes.

Roger Bacon assumed that al-Farghānī's cubit was a foot and a half long,[48] and in translating *Image du Monde* in 1480, William Caxton assumed that "xxv myles of Fraunce . . . is .l. englissh myle."[49]

Finally, we find al-Farghānī's version of the Ptolemaic System faithfully followed by Dante in his *Convivio*. In demonstrating the folly of trusting sense impressions alone, Dante wrote:[50]

> Thus we know that to most people the Sun appears to be a foot wide in diameter; and this is so utterly false that according to the investigation and discovery made by human reason with the attendant arts, the diameter of the Sun's body is five times that of the Earth's, and a half besides. For whereas the Earth has a diameter of six thousand five hundred miles, the diameter of the Sun, which when measured by sense impression seems to be a foot in extent, is thirty-five thousand, seven hundred and fifty miles.

In comparing the sphere of Mercury with dialectics, Dante assessed the planet's sizes as follows:[51]

> Mercury is the smallest star of the heaven, for the quantity of his diameter is not more than two hundred thirty-two miles, according to what Alfraganus assumes, who says that it is the twenty-eighth part of the diameter of the Earth, which is six thousand five hundred miles.

In the *Convivio* Dante also tells us that Venus's least distance is 167 times the Earth's half diameter.[52]

The *Divina Commedia* does not contain such explicit astronomical information, but we do learn in the ninth canto of *Paradiso* that the sphere of Venus is the highest sphere tainted by earthly inclinations, because this is the heaven "in which the shadow that your Earth casts comes to a point."[53] Now, Ptolemy and al-Farghānī gave the length of the shadow cone (*NX* in fig. 6, p. 17) as 268 e.r., [54] which is more than Venus's least distance but less than its greatest distance. The allegory is based on an astronomical fact.[55]

The Ptolemaic System, passed to the Latin West by Moslem authors of whom al-Farghānī was the most important in this instance, is found, then, at all levels of medieval literature. It, however, did not escape criticism in the more specialized philosophical and astronomical literature. In the thirteenth and fourteenth centuries a debate was carried on in European universities between those who preferred the homocentric spheres of Aristotle, modified by the Moslem astronomer al-Bitrūjī (Alpetragius, fl. ca. 1200), and those who insisted on Ptolemaic eccentrics and epicycles. In this debate, the variations in the magnitudes of the Moon, Venus, and Mars were cited as evidence against al-Bitrūjī by those who accepted Ptolemy's eccentrics and epicycles, such as Bernard of Verdun (late thirteenth century).[56] On the other hand, the fact that Ptolemy's models for these three bodies predicted much greater variations in their sizes or brightnesses than were, in fact, observed (the ratio of Mars's apogee to perigee distances is 7:1!) was cited by Henry of Hesse (1325–97) as a reason for rejecting Ptolemy and following al-Bitrūjī![57] Levi ben Gerson (1288–1344) also criticized Ptolemy for not paying attention to the variations in magnitudes of the planets, especially Mars. He chose to put both Mercury and Venus above the Sun and calculated Venus's apogee distance to be 8,971,112 e.r. Since his planetary theory is still not entirely understood, it is not known how he arrived at this remarkable distance, or his staggering distance of the fixed stars—159,651,513,380,944 e.r.[58]

With the possible exception of Levi ben Gerson, however, these criticisms were themselves based on the assumptions that the apparent diameters as they had been handed down were correct, that the principle of nesting spheres was valid, and that the planetary models in the *Almagest* contained the ratios of actual distances in the heavens. The criticisms were therefore squarely within the scholastic tradition of textual commentary.

The cosmic dimensions that originated with Ptolemy can thus be found on all levels of sophistication in medieval literature. This should not surprise us, for as an accepted part of the quadrivium, the regular mathematical subjects in the undergraduate curriculum, this scheme of sizes and distances was learned by virtually all university students. We may assume that it also frequently found its way, in full or in part, into the experience of the lay public.

The model of the cosmos in the minds of educated Europeans had a very detailed quantitative dimension. If all things had their (natural) places in the imagined universe of medieval man, there was nothing vague about the locations of these places. They were given very precisely by the Ptolemaic System which did not relinquish its hold on the minds of people until the seventeenth century.

* 5 *
Copernicus and Tycho

In 1500, thirteen and a half centuries after his death, Ptolemy still ruled the house of astronomy. The relevance of the *Almagest* had by no means diminished; in fact, with the renewed interest in original sources, Ptolemy's masterpiece had become more central to astronomy. The inadequacies of the tables and the calendar, as well as the problem of finding longitude at sea, directed the attention of astronomers back to the master himself. A century later astronomy had been transformed. New arrangements of the heavens had been added to Ptolemy's; precision instruments had changed the very practice of astronomy; and the authority of Ptolemy was beginning to erode.

This pivotal century, the age of Copernicus and Tycho Brahe, saw new cosmic dimensions take their place beside the old. These new dimensions were the by-product of the newly postulated arrangements of the heavenly bodies. As different as the heliocentric hypothesis of Copernicus and the geo-heliocentric hypothesis of Tycho Brahe were from the traditional geocentric system of Aristotle and Ptolemy, the break with traditional astronomy was not total by any means. This is illustrated by the ideas of Copernicus and Tycho concerning the sizes and distances of the heavenly bodies. Because of their agreement with Ptolemy on the order of magnitude of the solar distance, the sphere of Saturn and everything in it actually shrank by nearly half in their hands. With Tycho this was true for the entire cosmos! Except for the expansion of the sphere of the fixed stars necessitated by the heliocentric hypothesis, the new world systems were by themselves not enough to cause a truly radical reassessment of cosmic dimensions.

When Georg Peurbach (1423–61) and his pupil Johann Müller (1436–76), called Regiomontanus, set out to restore astronomy, a new translation of Ptolemy's *Almagest* directly from the Greek was the central aim of their quest. While the two scholars were not successful in their aim to produce a purified edition of Ptolemy's great work, they did place astronomical studies at European universities on a much higher level than they had ever been. Peurbach's *Theoricae Novae Planetarum*, first printed by Regiomontanus in about 1472, was more sophisticated than the medieval astronomy textbooks, and it became a standard text at most universities. The *Epitome of the Almagest*, which Peurbach started and Regiomontanus finished (first printed in 1496) was an important guide to the difficult *Almagest*. In this movement to upgrade astronomy and to focus astronomical study on the *Almagest* itself, commentaries such as al-Battānī's *De Motu Stellarum* were favored over elementary expositions such as al-Farghānī's *Differentie Scientie Astrorum*.[1]

Peurbach discussed Ptolemy's determination of solar distance by means of the eclipse diagram and gave al-Battānī's procedure as well, although he did not criticize either one.[2] We do not find al-Battānī's scheme of sizes and distances center-stage in the *Epitome*, but it is clearly in the wings. At the beginning of the ninth book where, following Ptolemy's plan, Regiomontanus discussed the order of the planets, he supported Ptolemy's argument that transits of Venus and Mercury could have occurred without having been noticed by citing al-Battānī's figure for Venus's apparent diameter. Since Venus's apparent diameter was only one-tenth the Sun's, Venus would only obscure one-hundredth of the Sun's disk, and such a small obscuration would not be visible. He also bolstered the Ptolemaic order of the planets by showing that Mercury and Venus just fit between the Moon and the Sun, but he was unaware that it was Ptolemy himself who had first made this argument and had invented the rationale—nesting spheres—for it. He used al-Battānī's figure of 1,070 e.r. for the Sun's least distance.[3]

Nicholas Copernicus (1473–1543) was part of the first generation to use the *Theoricae Novae Planetarum* and the *Epitome*, and he went on to produce the first astronomical treatise in the West comparable to the *Almagest*. In *De Revolutionibus Orbium Coelestium* (1543) Copernicus not only turned Ptolemaic astronomy on its head, but by rejecting the centrality of the Earth and asserting the centrality of the Sun he also brought about important changes in the scheme of sizes and distances that had been etched into the minds of students for centuries. The heliocentric hypothesis took the Moon out of the order of the planets and therefore destroyed the principle of the nesting spheres. Copernicus, however, added a new approach to the problem of sizes and distances, one which was more objective and in the long run more powerful than the nesting spheres procedure.

As we have seen, the order of the planets in Ptolemaic astronomy was a convention open to attack. Because the Moon eclipses all other heavenly bodies, it had to be the lowest. Arranging the planets and the Sun according to their increasing sidereal periods was a convenient, but not necessary, convention. Moreover, it did not solve the problem of how to arrange the bodies between the Moon and Mars, since Mercury, Venus, and the Sun all have the same sidereal period. The sequence above the Moon of Mercury-Venus-Sun-Mars-Jupiter-Saturn was accepted by astronomers on the basis of Ptolemy's authority alone; there was no compelling reason for it, and it was irrelevant to mathematical astronomy itself. For this reason, astronomers could, and sometimes did, disagree with Ptolemy's order of the planets while still staying well within the Ptolemaic paradigm.

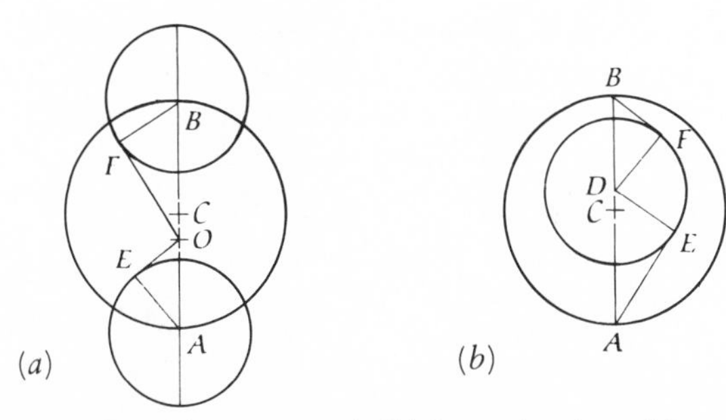

FIG. 8. Ptolemy's (a) and Copernicus's (b) determination of the parameters of the model of Venus

In the scheme outlined by Copernicus in book 1, chapter 10 of *De Revolutionibus*, the heliocentric periods of the planets increased with their heliocentric distances, if one assumed the order (starting from the Sun) of Mercury-Venus-Earth-Mars-Jupiter-Saturn. Mercury had the shortest period, Saturn the longest, and the Earth's period fell between those of Venus and Mars. This order also explained why Mercury and Venus appear bound to the Sun (Mercury within closer limits than Venus), while Mars, Jupiter, and Saturn do not. The heliocentric hypothesis, therefore, gave a more consistent picture of the order of the planets. Yet, up to this point this order was still only a convention.[4]

Consider, however, the case of Venus. Figure 8a shows the procedure used by Ptolemy to determine the parameters of the model, while figure 8b shows Copernicus's quite similar method. In figure 8a, O is the Earth, and Ptolemy had measured Venus's elongations from the Sun, $\angle EOA$ and $\angle FOB$ to be $47°20'$ and $44°48'$, respectively. Setting BC, the radius of the eccentric circle, equal to 60 parts, and knowing that the radius of the epicycle is constant, so that $EA = FB$, he found the eccentricity, CO, to be $1\frac{1}{4}$ parts, and the radius of the epicycle, EA or FB, $43\frac{1}{6}$ parts.[5] His results were not linked to those of the other planets in any way.

In figure 8b, the large circle with center C is the Earth's orbit, and the smaller circle with center D represents Venus's orbit. For Venus's greatest elongations from the Sun at apogee and perigee, $\angle DAE$ and $\angle DBF$, Copernicus used the same values as Ptolemy, $44\frac{4}{5}°$ and $47\frac{1}{3}°$. Since $\angle AED$ and $\angle BFD$ are right angles.

$DE = AD \sin 44\frac{4}{5}°$, and $DF = BD \sin 47\frac{1}{3}°$.

But $DE = DF$, so that $BD = 0.9593AD$.

Now, $AB = AD + BD = 1.9593AD = 2AC$.

So that $AC = 0.9797AD$.

Expressing $DE = DF$ in terms of AC,

$DE = 0.7193AC$.

Setting the radius of the Earth's orbit, AC, equal to unity, Venus's mean distance from the Sun was thus 0.7193.[6] From the mathematical model, therefore, Venus's distance from the Sun was found in terms of the Earth's

TABLE 4 Copernicus's Relative Planetary Distances

Body	Relative Distance in a.u.[a]		
	Least	Mean	Greatest
Mercury[b]	0.2627	0.3763	0.4519
Venus	0.7018	0.7193	0.7368
Earth	0.9678	1.0000	1.0322
Mars	1.3739	1.5198	1.6657
Jupiter	4.9802	5.2192	5.4582
Saturn	8.6522	9.1743	9.6963

[a]The distances of the planets are measured from the center of the Earth's orbit. The Earth's distance is measured from the true Sun.

[b]For Mercury's distances I have used the figures found by Maestlin. See Grafton, "Michael Maestlin's Account," p. 549.

distance from the Sun. By the same method Mercury's mean distance from the Sun could be found in terms of the Earth's,[7] while the relative mean heliocentric distances of Mars, Jupiter, and Saturn were found by slightly different geometrical procedures.[8]

The relative distances thus derived *directly from the planetary models* confirmed the order of the planets established by the criterion of increasing heliocentric periods. No new measurements were needed; these results could be obtained directly from the *Almagest.* For the superior planets, where the superimposed motion of the Earth is represented by the annual epicycles, the mean heliocentric planetary distances, expressed in terms of the radius of the Earth's orbit, are simply the ratios of deferent radii to epicycle radii. For the inferior planets the situation is reversed. All the parameters could be found in the *Almagest.*[9]

Copernicus did, however, construct new models and derived new parameters from observations. For each planet he gave the greatest, mean, and least distances from the Sun in terms of the Earth's mean orbital radius, the astronomical unit of heliocentric astronomy (hereafter a.u.). His results are given in table 4.[10]

Two things become immediately apparent when we tabulate Copernicus's results in this way: the Moon is missing from the sequence, and there are gaps between the greatest heliocentric distance of one planet and the least distance of the next one. This meant that the nesting spheres procedure, with the Moon as the starting point, was no longer possible. How, then, could one turn these relative distances into absolute ones?

Copernicus needed to know the absolute distance of any one body from the Sun, or the absolute distance between any two bodies on the table. He did not have observing instruments more accurate than Ptolemy's, so the direct determination of the parallax of any of these bodies was still out of the question. He could, however, determine the distance of the Moon.

Copernicus's lunar theory was a great improvement over Ptolemy's, for it did not predict great variations in the Moon's apparent diameter. Using

Ptolemy's procedure (see chap. 3), he found the Moon's greatest parallax to be of the order of 1°, substantially smaller than the value found by Ptolemy. His model predicted that the Moon's greatest and least distances at the quadratures were $68\frac{1}{3}$ e.r. and $52\frac{17}{60}$ e.r., respectively. The corresponding values at the syzygies were $65\frac{1}{2}$ e.r. and $55\frac{8}{60}$ e.r.[11]

Copernicus had two choices. He could use Aristarchus's method of lunar dichotomy to determine the ratio of solar to lunar distance, and then multiply the lunar distance by this ratio to determine the absolute distance of the Sun; or he could use the eclipse diagram to determine the distance of the Sun. Following the *Almagest*, he chose the latter method, but, as we shall see, Aristarchus's ratio was kept firmly in mind. Copernicus began by describing Ptolemy's procedure and taking notice of al-Battānī's changes in the parameters which led to a solar apogee distance of 1,146 e.r. (see p. 31), a value "which nevertheless cannot be coordinated in any way" with the parameters from which it was supposedly derived.[12] He then gave his own parameters:[13]

> With the intention of adjusting and correcting them, I put the apparent diameter of the sun at apogee = 31′40″, since it must now be somewhat bigger than before Ptolemy; [the apparent diameter] of the full or new moon when it is at its higher apse = 30′; the diameter of the shadow, where the moon passes through it, = $80\frac{3}{5}$′, because the ratio between them is recognized to be slightly bigger than 5:13, say 150:403; the entire sun at apogee is not covered by the moon, unless the latter's distance from the earth is less than 62 earth-radii; and the greatest distance from the earth to the moon in conjunction with or opposition to the sun = $65\frac{1}{2}$ earth-radii.

There are several noteworthy items in this statement. First, without any measurements Copernicus changed the apparent diameter of the Sun at apogee, because his solar theory predicted that the Sun's apogee distance had decreased since the time of Ptolemy.[14] Second, while Ptolemy had ruled out the possibility of an annular eclipse, Copernicus allowed it. Third, he determined the ratio of the shadow's diameter to the Moon's diameter, 403:150, at the Moon's greatest distance at syzygy, $65\frac{1}{2}$ e.r., and then assumed that this ratio would be the same at the lunar distance of 62 e.r., at which the Moon's apparent diameter was equal to that of the Sun's at apogee, 31′40″.

Referring once again to the now familiar eclipse diagram (figure 6, p. 17) from which we concluded that

$$ND = \frac{NH}{(\angle\beta/\angle\alpha + 1)\,NH\sin\angle\alpha - 1}$$

(eq. [6], p. 18), Copernicus's parameters were $NH = 62$ e.r.; $\angle\alpha = 15'50''$; $\angle\beta/\angle\alpha = 403/150$. From these parameters it follows that $ND =$

TABLE 5 Copernicus's Absolute Planetary Distances

Body	Relative Distance in e.r.[a]		
	Least	Mean	Greatest
Mercury	300	430	516
Venus	801	821	841
Earth	1,105	1,142	1,179
Mars	1,569	1,736	1,902
Jupiter	5,687	5,960	6,233
Saturn	9,881	10,477	11,073

[a]The distances of the planets are measured from the center of the Earth's orbit. The Earth's distance is measured from the true Sun.

1,175.5 e.r., but Copernicus put the Sun's apogee distance at 1,179 e.r.[15] Again, we are obviously not dealing with a solar distance computed from measured parameters: Copernicus carefully chose his parameters and made changes in the various drafts of *De Revolutionibus*, as Janice Henderson has shown.[16] It is to be expected that he wanted to find a solar distance not far from Ptolemy's; 1,179 e.r. was a thoroughly unobjectionable value, halfway between Ptolemy's and al-Battānī's solar apogee distances.

Copernicus found the solar eccentricity in 1543 to be 1 part in 31, somewhat less than Ptolemy's value of 1 part in 24. The Sun's mean distance was therefore 1,142 e.r. (corresponding to a solar parallax of almost exactly 3′), which happens to be almost exactly 19 times as great as the mean lunar distance, and this became the astronomical unit of heliocentric astronomy. All absolute distances within the sphere of Saturn were now known, but Copernicus did not convert his relative planetary distances to absolute distances. There was no need for this, and it would not have added anything to *De Revolutionibus*. For our purposes, however, it is useful to do so (see table 5).

Once again we see that the distances that followed from Copernicus's models left gaps between the spheres. Moreover, the spheres themselves were now much thinner. Whereas Ptolemy had needed spheres with thicknesses of 913 e.r. and 7,560 e.r. to accommodate Venus and Mars, Copernicus needed only 40 e.r. and 333 e.r. To satisfy the Aristotelian requirement that the cosmos be a plenum by postulating nesting spheres without spaces between them, a follower of Copernicus would have to fill the heavens with spheres of which only a minuscule portion was required to accommodate the eccentricities of the planets.[17] Worse, what was one to do with the staggering gap between Saturn and the fixed stars? Ptolemy's nesting spheres satisfied not only Aristotle's dictum that the cosmos is a plenum, but also his principle of economy: nature does nothing in vain. There were no empty spaces, and the spheres were exactly thick enough to accommodate the variations in geocentric distances of the planets, no thicker. Copernicus's spheres presented a dilemma. Either the cosmos was full and the spheres unnecessarily thick, or the spheres were just thick enough to accommodate the eccentricities and there were large spaces between them. In the long run the choice between plenitude and economy was decided in favor of the latter. Whereas for Aristotle and Ptolemy empty space by itself could at best only be an abstraction, for the followers of Copernicus it became a necessity and a reality.

Although the size of the Copernican cosmos had necessarily increased enormously, so that the distance to the fixed stars was "immense," table 5 reveals the surprising fact that *the sphere of Saturn and everything in it, except the Earth and the Sun, had shrunk by nearly half!* Even reckoned from the Earth, Saturn's greatest distance (at conjunction and aphelion, with the Earth also at aphelion) was now only 11,073 e.r. + 1,179 e.r. = 12,252 e.r. These changes would, of course, mean significant changes in the actual sizes of the planets, but Copernicus did not deal with this subject. His only reference to planetary sizes was in his discussion of the order of the planets, where, following Regiomontanus, he cited al-Battānī's (and Ptolemy's) value for Venus's apparent diameter to show that a transit of Venus would not be visible.[18] Copernicus did calculate the actual diameters and volumes of the Sun and Moon, but here there were no substantial changes. His solar distance and apparent diameter did not depart significantly from those of his predecessors, and although his lunar distances were significantly different, his estimate of the apparent lunar diameter differed in the opposite direction, so that his resulting absolute size of the Moon was almost the same as Ptolemy's.

Copernicus's contribution to the question of sizes and distances was enormous. By adding the simple, if controversial, postulate that the Sun, not the Earth, was the center of all planetary motions to the edifice of mathematical astronomy handed down from Ptolemy, he welded planetary theory and cosmology together in a unified system in which no further ad hoc hypotheses were needed. The order and relative distances of the planets followed directly from planetary theory, and for his followers the actual length of the astronomical unit was the only remaining problem. Contrasting his own system with that of his predecessors, Copernicus could write:[19]

> Having thus assumed the motions which I ascribe to the earth, . . . by long and intense study I finally found that if the motions of the other planets are correlated with the orbiting of the earth, and are computed for the revolution of each planet, not only do their phenomena follow therefrom but also the order and size[s] of all the planets and spheres, and heaven itself is so linked together that in no portion of it can anything be shifted without disrupting the remaining parts and the universe as a whole.

For the time being, however, the economy and harmony of the Copernican system were more than offset in the minds of astronomers by its disadvantages: the strangeness of a moving Earth and the abomination of such a huge universe with no apparent bounds—perhaps infinite. Yet Copernicus's achievement in mathematical astronomy was second only to Ptolemy's, and his planetary models became the basis for new tables. In preparing his *Tabulae Prutenicae* or *Prutenic Tables* (1551), Erasmus Reinhold (1511–53) addressed himself to the errors and inconsistencies in *De Revolutionibus*, and one of the obviously inadequate aspects was Copernicus's determination of the Sun's distance.

Reinhold's attempts to improve on Copernicus in this respect are preserved in a manuscript in his own hand that has been discovered in the twentieth century. In view of the extreme sensitivity of the eclipse method to small changes in the parameters, Reinhold's task was not easy. Janice Henderson has shown how, in an attempt to find a solar distance close to Ptolemy's from improved parameters, Reinhold arrived at values of 3,108 e.r., 1,301 e.r., and 1,187 e.r., rejecting each in turn. Finally, after many contortions and after introducing some inconsistencies of his own, he managed to obtain a solar apogee distance of 1,208 e.r., which he judged acceptably close to Ptolemy's 1,210 e.r.[20] This figure took its place in the astronomical literature beside the values of Ptolemy, al-Battānī, and Copernicus.[21] Of the four, Ptolemy's derivation was still the most consistent, because at least the result followed directly from the parameters he used.

* * *

Faced with the choice between the conflicting cosmic schemes of Ptolemy and Copernicus, astronomers naturally hoped to find a way to decide between the two. In principle there was a way. In the scheme of Ptolemy, Mercury and Venus were always below the Sun, while Mars was always above it. In the Copernican scheme, on the other hand, all three planets were sometimes nearer and sometimes farther than the Sun. Because of the proximity of the inferior planets to the Sun, observations at their near approaches to the Earth in the Copernican scheme were virtually impossible. But Mars provided an excellent opportunity for suitable observations. In the traditional order of the planets, Mars's least distance from the Earth was equal to the greatest solar distance, 1,260 e.r., according to Ptolemy himself, but 1,210 e.r. according to Copernicus's interpretation of Ptolemy's figure in the *Almagest*; 1,220 e.r. according to al-Farghānī; 1,146 e.r. according to al-Battānī; and 1,179 e.r. according to Copernicus himself. In the Copernican scheme, however, when Mars was in opposition to the Sun, that is, when Mars crossed the meridian at midnight, its distance from the Earth was only about ½ a.u., or about 750 e.r., according to Copernicus's measure. At a favorable opposition, when the Earth was near aphelion while Mars was near its perihelion, its distance from the Earth was even less. Since the Sun's parallax was about 3′, Mars's parallax at favorable oppositions should be more than 6′.[22] Therefore, if Mars's parallax at opposition (when in Ptolemy's scheme, too, it was closest to the Earth) could be measured, an objective decision between the systems of Ptolemy and Copernicus could be made.

Measuring angles accurately enough to detect differences as small as 6′ was not possible in Copernicus's lifetime. Not until the revolution in astronomical instrumentation and measuring techniques introduced by Tycho Brahe (1546–1601) and Landgraf Wilhelm IV of Hesse-Cassel (1532–92) did measurements of such small angles begin to come within the reach of astronomers under the best conditions. While still undecided about the merits of the two alternative schemes, Tycho attempted to measure the diurnal parallax of Mars during the favorable opposition that occurred in the winter of 1582–83. He gave two different accounts of these measurements, however. In 1584 he wrote to a correspondent that since he had measured Mars's parallax at opposition to be much less than the Sun's, Copernicus's scheme of the universe had to be wrong.[23] Beginning in 1587, however, he stated on several occasions that he had rejected the Ptolemaic system because he had measured Mars's parallax at opposition to be greater than the Sun's.[24] Since we know that Mars's horizontal parallax is at most 23″, a difference much too small to be measured by Tycho, how could he have measured a parallax at all?

From the researches of J. L. E. Dreyer (1852–1926), the editor of Tycho's works, it appears that Tycho decided that Mars had to be closer than the Sun when at opposition because its retrograde motion was then almost ½° per day. He was convinced that only if the planet were closer than the Sun when at opposition could it attain such a rapid apparent motion. In 1588 he gave this as his reason for rejecting the Ptolemaic system.[25]

Tycho decided on a geo-heliocentric system that gained great currency after the turn of the seventeenth century. Being geometrically equivalent to the Copernican system, this system predicted the same large parallax for Mars at opposition. The idea of measuring this parallax entered the literature with Tycho and was stressed by Kepler on several occasions. Later generations of astronomers with even more precise instruments would come back to this method. For the time being, as long as Mars's parallax at opposition was still swamped by the errors of the instruments, it nevertheless served to reduce the upper limit of the Sun's parallax to about half the instrumental error margin.

Tycho Brahe's geo-heliocentric system had all the advantages of the Copernican system without the two main disadvantages: the motions of the Earth and the great distance to the fixed stars. In Tycho's system, too, the order of the planets and their relative distances from the Sun followed directly from the models, and these distances were the same as in the Copernican system. Tycho did not, however, as we might expect, redetermine the parameters necessary to calculate the solar distance by means of the eclipse diagram. Instead, he chose a mean solar distance of 1,150 e.r., giving as his justification that it was halfway between Copernicus's value of

TABLE 6 Tycho Brahe's Scheme of Sizes and Distances

Body	Mean Distance (in e.r.)	Apparent Diameter at Mean Distance	Actual Diameter (Earth = 1)	Volume (Earth = 1)
Moon	60	33′	49/176 ≈ 2/7	ca. 1/40
Sun	1,150	31′	5 14/75	ca. 140
Mercury	1,150	2 1/6′	3/8	ca. 1/19
Venus	1,150	3 1/4′	6/11	6/37
Mars	1,745	1 2/3′	127/300	ca. 1/13
Jupiter	3,990[a]	2 3/4′	12/5	ca. 14
Saturn	10,550	1′50″	31/11	ca. 22
New star of 1572	14,000	3 1/2′	7 1/8	ca. 360
Stars				
M1	14,000	2′	4 1/3	ca. 68
M2	14,000	1 1/2′	3 1/18	ca. 28 1/2
M3	14,000	1 1/12′	20/9	ca. 11
M4	14,000	3/4′	55/36 ≈ 3/2	ca. 27/8
M5	14,000	1/2′	50/49	ca. 1 1/18
M6	14,000	1/3′	15/22	ca. 1/3

[a]Jupiter's apparent diameter was measured at *least* distance, but its actual diameter was calculated assuming that the apparent diameter is 2 3/4′ at mean distance, about 6,000 e.r.

1,142 e.r. and Ptolemy's supposed value of 1,160 e.r., and that it was, at any rate, very close to "that mystical number," 576, which would express the same measure in terrestrial diameters![26] Tycho's scheme of absolute distances was, therefore, hardly different from Copernicus's, except, of course, that all measurements were referred to the stationary Earth, not the Sun. There was one major difference, however. Since Tycho rejected the Earth's annual motion, he had no need to remove the fixed stars to a distance great enough to explain the lack of a detectable annual stellar parallax. And since by his measure Saturn's greatest distance from the Earth was 12,300 e.r., Tycho put the sphere of the fixed stars immediately above it, at 14,000 e.r.[27] Now, therefore, *the entire cosmos had shrunk by a third from the old measure.*

Up to this point no one had questioned the planetary apparent diameters going back to Ptolemy and Hipparchus. Tycho now made his own estimates of these measures, but he was careful to point out that he claimed no great accuracy in this endeavor. His results were very close to Ptolemy's, but since his distances were different, the actual sizes of the planets came out differently in most cases (see table 6).[28]

Comparing these sizes and distances with the traditional schemes of Ptolemy, al-Farghānī, and al-Battānī, we see that whereas the inferior planets—Mercury and Venus—had become larger, the superior planets—Mars, Jupiter, and Saturn—had shrunk, while the Sun and Moon had remained about the same (although Tycho's much improved lunar theory resulted in much smaller variations in the Moon's distance and therefore its size). Since Tycho's system was geometrically equivalent to the Copernican system, these changes were welcome news to Copernicans. The Sun was now almost 7 times as large in volume as its nearest rival in the solar system, Saturn, as befitted him who "governs the family of planets revolving around it."[29] Moreover, Mercury, which had been smaller than the Moon, was now larger, and a primary planet ought to be larger than a secondary one. Within the orb of Saturn, therefore, Tycho's scheme of sizes lent support to the Copernican notion of planetary hierarchy.

Above Saturn things were different, however. Tycho had roughly confirmed the earlier measures of the apparent diameters of the fixed stars. Since his distance of the fixed stars was smaller than the traditional measure, their actual sizes had diminished somewhat. A Copernican, however, was forced to put the fixed stars at an enormous distance from the Sun and Earth. Given the apparent diameters handed down from Hipparchus and Ptolemy and now confirmed by Tycho, what would such a distance mean for the actual sizes of the fixed stars? Only when we realize the canonical status of the apparent diameters, to which Tycho had also lent his enormous authority, can we appreciate the force of his argument against the Copernican hypothesis.

Tycho argued that with his instruments, more accurate than any instruments in history, he had not been able to detect any annual motion of the fixed stars that would be a reflection of the Earth's motion around the Sun. Knowing the accuracy of his instruments, Tycho therefore set 1′ as the upper limit of annual stellar parallax. This meant that in a heliocentric cosmos, for an imaginary observer on one of the fixed stars, the diameter of the Earth's annual circle around the Sun would subtend an angle of at most 1′. From this it followed that the distance to the fixed stars was at least 2,300/sin 1′ e.r. Tycho simply stated that in the Copernican scheme the fixed stars had to be at least 700 times as far away from the Sun as Saturn is, that is, about 7,850,000 e.r. using Tycho's solar distance of 1,150 e.r.

Moreover, if seen from the Earth a fixed star of the third magnitude had an apparent diameter of 1′ (see table 6), while seen from the fixed stars the Earth's orbit had an apparent diameter of 1′, then a fixed star of the third magnitude had an actual diameter as large as the entire orbit of the Earth: 2,284 e.r. according to the measure of Copernicus; 2,300 e.r. according to Tycho's measure.[30] Thus a third magnitude fixed star would have a diameter more than 200 times as large as the Sun's, and a large fixed star such as Sirius or Lyra, which had an apparent diameter of 2¼′ according to Tycho,[31] would be comparably larger.

Tycho's logic was impeccable; his measurements above reproach. A Copernican simply had to accept the results of this argument. Thus, when in 1589 Tycho maintained in a letter to Christoph Rothmann (ca. 1550–1605) that such distances and sizes were absurd,[32] Rothmann replied:[33]

> Moreover, concerning the annual motion, why would it seem unlikely to me that the space from the Sun to Saturn is contained many times between Saturn and the distance of the fixed stars? Or what absurdity follows if a star of the third magnitude equals the entire annual orb? Does it perhaps conflict with divine Will, or is it impossible to divine Nature, or does it not agree with the infinity of Nature? If you want to deduce something absurd from this, it is entirely up to you to demonstrate this. The absurdity of things, which at first glance appear so to the multitude, cannot be so easily demonstrated. Indeed, divine Wisdom and Majesty is much greater, and whatever size you concede to the Vastness and Magnitude of the World, it will still have no measure compared to the infinite Creator. The greater the King, the larger and more spacious a palace he deems fitting to his Majesty. And what will you think of God?

Although as a Copernican Rothmann was willing, by necessity, to let the evidence guide him, he had to appeal to a higher authority to balance the weight of tradition. Tycho's objection was based on a notion of cosmic sizes and distances that was deeply etched into the collective consciousness of educated Europeans. Therefore his objection was a very serious one; no answer was in sight.

In the meantime, the Italian astronomer Giovanni Antonio Magini (1555–1617), a defender of Ptolemaic astronomy who nevertheless incorporated Copernicus's improvements of planetary theory in his work, strengthened the hand of the anti-Copernicans even more. In his *Novae Coelestium Orbium Theoricae Congruentes cum Observationibus N. Copernici* (1589), he insisted on much larger apparent diameters for the planets and fixed stars, and he gave apogee and perigee values: 5′ and 7′ for Saturn, 8′ and 11′ for Jupiter, 6′ and 10′ for Mars, 9′ and 12′ for Venus, and 5′ and 9′ for Mercury. According to Magini, fixed stars of the first magnitude had apparent diameters of no less than 10′, while the remaining magnitudes showed disks of 5′ or 6′, 4′, 3′, 2′, and 1′, respectively.[34] Now even stars of the sixth magnitude would have diameters equal to the Earth's annual orbit if one allowed the Earth an annual motion. Magini did not depart from the traditional figure for the solar distance,[35] and thus, by implication, he made the actual sizes of the planets and fixed stars much larger than they had ever been. Although he held the chair of astronomy at the University of Bologna (a position for which he had been preferred over Galileo) and his works were popular, Magini's notions about cosmic sizes were too eccentric to be influential. Yet, they have some historical importance because Galileo used them to demonstrate the absurdity of pretelescopic "measurements" of the sizes of planets and fixed stars.

On the eve of telescopic astronomy, then, there were a variety of schemes of sizes and distances from which to choose. These fell into two families: Ptolemaic and Copernican / Tychonic. For the orthodox astronomer the works of al-Farghānī and al-Battānī, both available in print, as well as university textbooks presented schemes that were almost identical. The followers of Copernicus and Tycho could only quarrel with each other about the sizes and distances of the fixed stars. The two families had in common a close agreement on the lunar and the all-important solar distance, as well as all apparent diameters. The only disagreement was on planetary and stellar distances. Yet even in the case of the planets (although, e.g., al-Farghānī's greatest distance of Saturn, 20,110 e.r., was substantially different from Tycho's figure of 12,300 e.r.) there was a comfortable underlying agreement on the order of magnitude of planetary distances. For this reason it is entirely reasonable to see Tycho's scheme of sizes and distances as a variation on the traditional scheme, a variation that made no one uncomfortable. Only the sizes and distances of the fixed stars that followed from the Copernican hypothesis presented a radical break with tradition.

Although astronomers could quibble with each other, their notions of how big heavenly bodies were and how far away they were had not changed markedly since Ptolemy. When, in his last (posthumously printed) edition of his commentary on the *Sphere* of Sacrobosco (1612), Christopher Clavius (1537–1612) presented a table of sizes and distances virtually identical to al-Farghānī's,[36] he was expressing a consensus that still reigned, a quantitative picture of the cosmos that was built into the minds of all students during their undergraduate training. In a cosmos still predicated on the concept of place, every place was quantitatively known with great precision by all university graduates who could remember the numbers. Those who could not nevertheless had a good idea of the orders of magnitude involved.

This quantitative world picture was, therefore, an important part of the collective consciousness of educated Europe, and astronomical theories were judged in its light. Only the logic of his own system allowed, or rather forced, Copernicus to transcend it in the case of the fixed stars, and those huge distances made his system all the more objectionable to others. Tycho's objection to Copernicus, on the other hand, appealed to the shared traditional expectations of how large heavenly bodies ought to be.

The conceptual strain presented by the Copernican hypothesis could only be resolved when his proposed stellar distances had become somewhat less offensive with time and when radically new evidence forced astronomers to change their expectations of cosmic dimensions by orders of magnitude.

* 6 *
Young Kepler

Despite a tension between Aristotle's cosmology and Ptolemy's mathematical astronomy, the Ptolemaic world system was firmly imbedded in Aristotle's cosmology and physics. One supported the other. Copernicus's world system, on the other hand, was not supported by any established system of physics. The price for believing that the Sun was actually the center of the cosmos was the loss of all the explanatory power of Aristotelian physics. The Copernican world system therefore needed a supporting physics and even metaphysics of its own, and in the absence of such support the heliocentric theory tended to be used as a mathematical hypothesis for purposes of calculation only. For half a century after its publication the Copernican world system attracted few true believers, and little progress was made toward making the system more attractive.

The first major contribution to Copernican astronomical theory came at the hands of Johannes Kepler (1571–1630), perhaps the most creative mind ever to adorn astronomy. From the very beginning of his astronomical career in the 1590s, Kepler squarely faced the problem posed by the Copernican hypothesis: to give causal, physical explanations of how bodies could be arranged and move in the way postulated by Copernicus. This meant introducing physics into the heavens, thus obliterating the traditional separation between mathematical astronomy and philosophy, between mere mathematical hypothesis and reality. Moreover, convinced that God had constructed the world from a mathematical blueprint, Kepler looked for mathematical relationships that expressed the hidden harmony in the Creator's plan.

Kepler's ambitious plan to reform astronomy thus departed radically from traditional notions about what constituted the proper study of the heavens. Because of his unorthodox approach, he was able to formulate an entirely new sort of scheme of cosmic dimensions. If, in the long run, this new scheme was not successful, it was nevertheless important because, in departing entirely from the traditional approach to cosmic dimensions, Kepler helped lay it to rest and prepare the ground for thoroughly new and ultimately successful efforts.

As a mathematics teacher at Graz, young Kepler was bothered by a problem created by Copernicus: the huge empty spaces left between the thin planetary spheres. Consider again table 4 (p. 44). According to Copernicus's models, Jupiter's least distance was 4.9802 a.u., and its greatest distance 5.4582 a.u. The corresponding measures for Saturn were 8.6522 a.u. and 9.6963 a.u. Jupiter's sphere was therefore less than ½ a.u. thick, Saturn's sphere was slightly over 1 a.u. thick, and between these spheres there was a space of over 3 a.u. How did one explain such a huge space between two such thin spheres? Was there an underlying rationale that explained those particular numbers? For that matter, why was the Sun surrounded by six planets and not twenty or a hundred?[1] For Kepler these questions required mathematical answers. Was there a mathematical construction that could, for example, produce the ratio 5.4582/8.6522? If there were no crystalline spheres, could one at least "fill" space with mathematical figures or ratios? Was there a mathematical demonstration for the necessity of six planets? He spent the rest of his life pursuing the answers to these questions.

Kepler's first solution came to him in 1595. It had been known since Plato's time that there could be only five regular solids in three dimensions: the regular tetrahedron, octahedron, and icosahedron with four, eight, and twenty equilateral triangles as faces, the cube with six squares as faces, and the dodecahedron with twelve regular pentagons as faces.[2] If these figures could be fitted between the Copernican planetary spheres, they would explain why there were six and only six planets. Was it possible to find a sequence so that each planetary sphere exactly touched the faces of a surrounding regular solid, the vertices of which in turn exactly touched the inside of the next larger planetary sphere? The ratios of the radii of the inscribed spheres to the circumscribed spheres were as follows:[3]

tetrahedron	0.333
octahedron	0.577
icosahedron	0.795
cube	0.577
dodecahedron	0.795

From his teacher, Michael Maestlin (1550–1631), Kepler knew the following relative planetary distances, derived from *De Revolutionibus*:[4]

Saturn:	greatest	9.6964
	least	8.6564
Jupiter:	greatest	5.4581
	least	4.9803
Mars:	greatest	1.6656
	least	1.3739
Earth:	greatest	1.0417
	least	0.9583
Venus:	greatest	0.7611
	least	0.6778
Mercury:	greatest	0.4900
	least	0.3006

TABLE 7 Kepler's First Effort to Account for Celestial Distances by Means of the Five Regular Solids

If the Radius of the Inner Surface Is 1000		Then the Radius of the Outer Surface of the Next Smaller Sphere Is: Computed	Copernicus
Saturn	Jupiter	577	635
Jupiter	Mars	333	333
Mars	Earth	795	757 (801)[a]
Earth	Venus	795	794 (847)[a]
Venus	Mercury	577 or 707	723

[a]The numbers in parentheses result when the thickness of the Earth's sphere is increased sufficiently to accommodate the orbit of the moon.

The ratios between the greatest distance of one planet and the least distance of the next planet were, then,[5]

Jupiter/Saturn	0.631
Mars/Jupiter	0.334
Earth/Mars	0.758
Venus/Earth	0.794
Mercury/Venus	0.723

In two cases there was an obvious fit. The tetrahedron would fit almost perfectly in the large space between Mars and Jupiter, while either the dodecahedron or the icosahedron would just about exactly fill the space between Venus and the Earth. Kepler chose the icosahedron. The three remaining solids did not provide very good fits with the remaining spaces. Mercury was especially problematical, and in putting the octahedron between Mercury and Venus, Kepler made the greatest distance of Mercury coincide with the radius of the circle inscribed in the square formed by the four middle sides of the octahedron, which gave the ratio 0.707 instead of 0.577. He thus arrived at the first approximation, shown in table 7.[6]

Kepler had thus obtained good results for Mars and Venus but rather poor results for Jupiter, the Earth, and especially Mercury. Increasing the thickness of the Earth's sphere to accommodate the Moon's orbit did not change the figures computed from the regular solids since the ratios are independent of the thicknesses of the spheres. This would, however, change the figures for the Earth and Venus according to Copernicus to the values given in parentheses (table 7). As can be seen, the overall quality of the fit was not affected by this change.

Although the agreement was not as good as Kepler would have hoped, he nevertheless judged his results acceptable (except for the case of Saturn/Jupiter).[7] His standards of comparison, the relative distances according to Copernicus, were, however, by no means perfect either. When he acquainted Maestlin with his discovery, Maestlin expressed surprise that Kepler had uncritically taken planetary distances from Maestlin's table, copied directly from Copernicus. Surely, for a work such as this, these

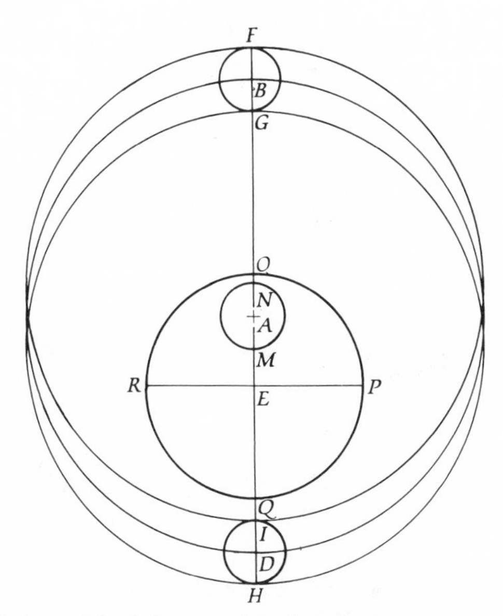

FIG. 9. Maestlin's model of the superior planets.
E is the Sun; *OPQR* is the orbit of the Earth; *NM* is the movable eccentric on which the deferent is centered. At apogee the planet is at *G*, not at *F*; at perigee it is at *H*, not at *I*.

distances should be recalculated from the more accurate *Prutenic Tables* of Reinhold.[8] Maestlin sent him a little tract in which he had performed these calculations.[9] In the ensuing exchange of letters, Kepler and Maestlin not only worked out tricky mathematical problems such as the distances derived from the model of Mercury,[10] but they also discussed the concept of a sphere and made Copernican astronomy truly heliocentric.

Maestlin wished to make corrections in Kepler's distances. Kepler had confused the greatest and least distances of the planets from the center of the Earth's orbit with the greatest and least radii of their *spheres*, and Maestlin wrote therefore:[11]

> For example, when Saturn is in the apogee of its eccentric, it is [9.70] distant from the center of the great sphere where the semidiameter of the great sphere is [1.0]. But Saturn itself is then in the perigee of its epicycle; therefore, the apogee of the epicycle is then much farther [from the center of the great sphere]. The opposite holds if it is in the perigee of its eccentric. And if we use an eccentric on an eccentric, or an epicycle on an epicycle, in place of an epicycle on an eccentric, the result will be the same. But your numbers seem to take the greatest and least distances of the planets from the center of the great sphere as the size of the spheres, although that is in fact much greater.

As can be seen in figure 9, at apogee Saturn's distance from the Sun is *EG*, but the radius of its sphere which can accommodate its entire epicycle is then EF. Likewise, at perigee, Saturn's distance from the Sun is EH, but the inside radius of the sphere which accommodates its epicycle is only EI. Should the thickness of Saturn's sphere be reckoned as $EF - EI$, or merely as $EG - EH$? Maestlin, who still adhered to a concept of celestial spheres closely related to the earlier notion (although his spheres were no longer crystalline), believed that the first measure was the correct one. Kepler, however, did not share his teacher's concept of celestial spheres and rejected the proposed correction:[12]

> . . . even if each planet truly has this epicycle, nevertheless, so long as it does not ascend higher than my values state, that is enough for me. I say in my book that I doubt whether there are spheres at all.

And, indeed, in the *Mysterium Cosmographicum*, which came off the press in March 1597, Kepler argued against the need for, indeed, the existence of planetary spheres.[13]

Maestlin's more accurate distances, derived from the *Prutenic Tables*, varied very little from those given by Copernicus. But in constructing his mathematical model which unified for the first time all the planetary orbs or spheres into one single *system*, Kepler discovered another problem. Both Copernicus's and Maestlin's relative distances were measured from the center of the Earth's orbit. This point (which is removed from the Sun by the amount of the Earth's eccentricity) was therefore the center of the system of spheres and regular solids. It was also the center of the Earth's sphere. Since Copernicus had made the Earth's orbit exactly circular (albeit not exactly centered on the Sun), this meant that in Kepler's construction the Earth's sphere had no eccentricity and no thickness. In his first table, which only gave otherwise independent ratios, this was no problem. If, however, one wished to calculate the relative distances from the center predicted by Kepler's system, problems arose. Kepler saw that the only way to construct his system was to make the Sun the true center. Again he turned to Maestlin, who made the appropriate calculations and, perhaps for the first time in the history of astronomy, produced a figure in which the eccentricities and epicycles of all the planets, including the Earth, were shown at the same time. This departure made mathematical astronomy Sun-centered, strictly speaking.[14]

Kepler now produced a table (see table 8) in which he compared relative planetary distances measured from the center of the Earth's orbit (1) and from the Sun itself (2) with the distances predicted by his scheme of regular solids, ignoring the Moon's orbit (3) and taking that orbit into account (4). In calculating the distances predicted by the regular solids, he used the ratio of aphelion to perihelion distance for each planet found in column (2).[15]

Comparing Kepler's predicted values (3) with Maestlin's heliocentric distances (2), we see that the agreement is, in fact, rather good, except for

TABLE 8 Kepler's Actual and Predicted Planetary Distances in *Mysterium Cosmographicum*

Planet		Relative Distance in a.u.[a] (1) According to Copernicus & Reinhold	(2) True Helio-centric	(3) Predicted by Regular Solids	(4) Prediction with Moon Included
Saturn	max.	9.7000	9.9875	10.5989	11.3044
	min.	8.6500	8.3417	8.8522	9.4406
Jupiter	max.	5.4581	5.4925	5.1108	5.4506
	min.	4.9803	4.9994	4.6522	4.9606
Mars	max.	1.6656	1.6478	1.5506	1.6536
	min.	1.3739	1.3931	1.3108	1.3978
Earth	max.	1.0000	1.0417	1.0417	1.1017
	min.	1.0000	0.9583	0.9583	0.8983
Venus	max.	0.7611	0.7414	0.7614	0.7139
	min.	0.6778	0.6964	0.7153	0.6706
Mercury	max.	0.4900	0.4886	0.5058	0.4742
	min.	0.3006	0.2333	0.2333	0.2186
Sun	max.	0.0417[b]	—	—	—
	min.	0.0323[b]	—	—	—

[a]Kepler gives the distances in sexagesimal numbers.

[b]These are the maximum and minimum values for the solar eccentricity according to Copernicus (*De Revolutionibus,* chaps. 3, 16).

the cases of Jupiter and Saturn. Adding the Moon's orbit to the Earth's sphere (4) made the agreement a bit better for Jupiter but worse for Saturn. Could the agreement, such as it was, be coincidental?

Kepler now considered the periods of the planets, and here fixing the center of his system in the Sun itself dovetailed with his introduction of physics into astronomy. Obviously, the heliocentric periods of the planets do not vary directly as their distances from the Sun.[16] Saturn is about 9 times as far from the Sun as the Earth is, but its period is almost 30 times as long as the Earth's. Kepler introduced the notion of an *anima motrix:* rejecting the possibility that each planet had an individual moving soul, he chose one single moving soul with its seat in the Sun itself. In emanating outward, the power of this soul declined with increasing distance, so the farther a planet was from the Sun, the smaller the power was that pushed it along in its path.[17]

Kepler argued that a more remote planet takes longer to complete its journey around the Sun for two reasons: the circumference of its circle is greater, and the power pushing it along this circumference is smaller. Since both factors vary with the radius of the circle, they each account for half the increase. Thus, since Mercury's period is about 88 days and Venus's about 224½ days—a difference of 136½ days—only half this difference, 68⅓ days, is to be ascribed to the difference between their distances from the Sun. Accordingly, the radii of the orbits of Mercury and Venus should

TABLE 9 Kepler's Attempt to Account for the Distances of the Planets by the Argument from Their Periods

Planets	Ratio of Mean Distances	
	According to Kepler's Scheme	According to Copernicus
Jupiter:Saturn	0.574	0.572
Mars:Jupiter	0.274	0.290
Earth:Mars	0.694	0.658
Venus:Earth	0.762	0.719
Mercury:Venus	0.563	0.500

be to each other as 88:(88 + 68⅓) or 88:156⅓.[18] Kepler calculated the ratios of the periods and compared them with the ratios from Copernicus (see table 9).[19] Again, the agreement was, if not perfect, good enough to reinforce Kepler's faith in intelligent design.

When *Mysterium Cosmographicum* was finished, Maestlin saw it through the press in Tübingen, and he took the opportunity to include an edition of the *Narratio Prima* (1540) of Georg Joachim Rheticus (1514–76) which had become scarce. He also included an appendix, "On the Dimensions of the Heavenly Circles and Spheres, according to the *Prutenic Tables*, after the Theory of Nicolaus Copernicus," a polished and slightly altered version of Maestlin's response to Kepler's first call for help with the planetary distances.[20] Here, Maestlin gave a succinct account of the relevant parts of Copernican planetary theory and derived the greatest and least distances of each planet from the center of the Earth's orbit. His reconstruction of Reinhold's greatest solar distance, 1,208 e.r., from the parameters in the *Prutenic Tables*[21] allowed the reader to convert relative distances to absolute ones. It was an important, indeed necessary, appendix to the *Mysterium Cosmographicum*, and because of it and the new edition of the *Narratio Prima*, Kepler's first publication enjoyed an enhanced popularity.

When, largely as a result of the *Mysterium Cosmographicum*, Kepler joined the staff of Tycho Brahe in Prague in 1600, he obtained access to the most accurate observations ever produced, and against these he continued to test his mathematical and physical ideas. While a few years earlier he had accepted Maestlin's solar distance as given in the appendix of the *Mysterium* without comment, he now began to look into this matter more deeply. By the end of 1605 Kepler had completed his *Astronomia pars Optica* as well as his *Astronomia Nova*.[22] In the former, published in 1604, he examined the problem of parallax in general terms and promised the reader a work entitled "Hipparchus" in which he would deal with the problem of solar and lunar parallax in eclipse theory.[23] In *Astronomia Nova*, not published until 1609, he examined the problem of Mars's parallax in detail.

Tycho and his assistants had tried to measure Mars's diurnal parallax at opposition over several years. However, as mentioned in chapter 4, Tycho had issued conflicting statements on the results. After examining Tycho's records, Kepler thought he could clear away some of the confusion. From the observations themselves, no parallax, or only a very small one, could be deduced. Tycho had asked his assistants to measure Mars's parallax from the morning and evening measurements of its positions compared to the fixed stars—its so-called diurnal parallax. The assistants, Kepler argued, had apparently misunderstood their orders and had calculated the planet's parallax at that position according to the Copernican geometry and Copernicus's solar distance. Of course, Mars's parallax came out to be considerable: about 6′.[24] In the twentieth century, however, Dreyer has shown Kepler's explanation to be mistaken, for a manuscript in Tycho's own hand shows the same calculation as his assistants made.[25] As discussed earlier (p. 49), Mars's velocity at opposition apparently led Tycho to reject the Ptolemaic scheme of the heavens.

A careful analysis of Tycho's observations, as well as some of his own, led Kepler to conclude that Mars's diurnal parallaxes were not detectable even with Tycho's accurate instruments. The planet's greatest parallax was certainly no greater than 4′, an amount that could be ascribed to the inaccuracy of the instruments, and was in all likelihood very small.[26]

Since Mars's parallax at opposition was greater than that of the Sun and yet could not be determined, the investigation of the Sun's distance and parallax by means of the eclipse diagram was an even more uncertain procedure. Kepler was certain, however, that the Sun was more than 230 e.r. and less than an infinite distance removed from the Earth. He claimed to have established this lower limit in his *Mysterium*, although it does not appear there.[27] We may assume, however, that Kepler arrived at it by postulating that the Sun could not be smaller than the Earth. Since the Sun's angular radius is about 15′, if its distance were less than cotan 15′, or about 230 e.r., the Sun would be smaller than the Earth. Moreover, from his study of eclipses Kepler claimed he had been able to narrow the range of solar distances further to between 700 e.r. and 2,000 e.r., as he would show in his "Hipparchus."[28]

At Kepler's death, his "Hipparchus" remained unpublished, and the fragments printed for the first time by C. Frisch in the nineteenth century date from Kepler's last decade, when he had rejected the eclipse diagram as a tool for finding the solar distance and parallax.[29] From the tract it is clear that Kepler was at least the equal of Ptolemy in the use of this diagram. In fact, he simplified the procedure considerably. Consider again the eclipse diagram (fig. 10). Kepler showed that $\angle NXR = \angle DNC - \angle NCM$, and $\angle NXR = \angle MRN - \angle XNR$. From this he produced the equation:

$$\angle DNC + \angle XNR = \angle MRN + \angle NCM.$$

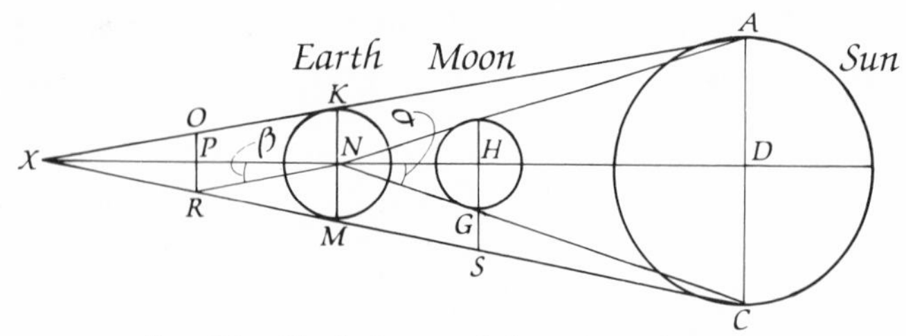

FIG. 10. Kepler's use of the eclipse diagram

In other words, the sum of the Sun's angular radius and the shadow's angular radius equals the sum of the horizontal parallaxes of the Sun and Moon.[30]

Late in his life, Kepler did not use this equation to determine solar parallax. Like Hipparchus, after whom he named the tract, he *assumed* a solar parallax, but where Hipparchus had then derived a lunar distance, Kepler already knew the lunar distance. He used the equation to calculate the angular radius of the shadow, $\angle XNR$, so that he could compute the extent and duration of a lunar eclipse.[31] We may surmise, however, that early in his career he also used the equation to determine solar distances.

The beginning of Kepler's disillusionment with the eclipse diagram for determining solar distance can be dated to about 1606. Two years earlier a new star had appeared in the heavens, and, after finishing the *Astronomia Nova*, Kepler wrote *De Stella Nova in Pede Serpentarii*, printed in 1606. In it he discussed the size and distance of the new star as well as the distribution of fixed stars in the heavens. The problem of the distance was approached in typical Keplerian, a priori fashion. Three things are required for motion: a mover, things moved, and a place. In this case the Sun is the mover, the planets are the things moved, and the outermost sphere of the fixed stars is the place. Now, so that the physical could be expressed in mathematical terms, Kepler postulated that the things moved are the mean proportional between the mover and the place. He made the sphere of Saturn the mean proportional between the body of the Sun and the sphere of the fixed stars. But how big was the Sun? Or rather, what was its distance?[32]

Without any explanation or reference to his "Hipparchus," Kepler chose a solar distance of 1,432 e.r., which corresponds to a solar parallax of $2'24''$. According to him the Sun's diameter ($1{,}432 \times \tan 15' = 6.25$ e.r.) was about 6 e.r. and the radius of the sphere of Saturn was about 10 a.u. or 14,320 e.r., clearly convenient estimates. The radius of the sphere of the fixed stars, R, could thus be found from the relationship $6{:}14{,}320 = 14{,}320{:}R$, so that R was 34,177,066⅔ e.r.[33] Whereas Copernicus had only gone as far as saying that the distance of the fixed stars was immense, Kepler had derived an actual distance, and it was indeed immense. For

Kepler the radius of the cosmos was almost 2,000 times as large as it had been for Ptolemy and his successors up to Copernicus.

The apparent magnitudes of the heavenly bodies were not a matter of mathematics but rather one of observation and measurement, and here Kepler did not disagree with Tycho and his predecessors. He argued that if the new star had an apparent magnitude only as large as Sirius's, 4′, it would be larger than the entire sphere of Saturn![34] (Its diameter would be 34,177,067 tan 4′ = 39,767 e.r., while Kepler's value for the diameter of Saturn's sphere was 28,640 e.r.) Obviously, in 1606 Kepler had no answer to Tycho's objection to the Copernican hypothesis (see p. 51). He did, however, have an answer to Giordano Bruno (1548–1600).

Bruno had argued that the universe was infinite and contained innumerable suns surrounded by planets. Kepler characterized Bruno's position as follows:[35]

> . . . Bruno made the world so infinite that [he posits] as many worlds as there are fixed stars. And he made this our region of the movable [planets] one of the innumerable worlds scarcely distinct from the others which surround it; so that to somebody on the Dog Star . . . the world would appear from there just as the fixed stars appear to us from our world. Thus according to them, the new star was a new world.

Kepler attacked the supposition that the world ought to look the same from every vantage point with an argument based on sizes and distances. Consider the three second-magnitude stars that make up the belt of the constellation Orion. The middle one was, according to Kepler, removed from the outer two by 81′ or less, and each showed an apparent disk of at least 2′.[36] Under Bruno's supposition that all fixed stars are of the same size, so that their magnitude is a function of their distance only, they are at the same distance from us. Kepler argued:[37]

> . . . if they were placed on the same spherical surface of which we are the center, the eye located on one of them would see the other as having the angular magnitude of about 2¾°; [a magnitude] that for us on the earth would not be occupied by five suns placed in line and touching each other.

For an observer on one of these stars, two huge suns would appear in the sky, not to speak of the many smaller stars between these larger stars in Orion's belt. Obviously, the world looks very different to an imagined observer among the fixed stars. The Sun, therefore, occupies a singular and, by implication, central position in the cosmos.[38] Until the end of his life, Kepler continued to believe that the fixed stars were arranged in a spherical shell, although he was to change his mind about the actual sizes of the stars.

Although in his examination of planetary and stellar distances Kepler was much less constrained by tradition than his predecessors and contemporaries, when it came to planetary and stellar sizes his ideas were, at this stage of his career, still entirely conventional. As we have seen, he thought that the apparent diameters of the three large stars in Orion's belt were at least 2′. In 1607 these ideas about planetary sizes led him astray.

In a discussion of occultations of planets and fixed stars in his *Astronomia pars Optica*, Kepler had reproduced a report by Einhard in his *Annales Regni Francorum*: "On 15 April the star of Mercury appeared on the Sun as a small black spot . . . which was seen by us for eight days."[39] Kepler agreed that his ninth-century predecessors had actually seen Mercury on the Sun, but such an event could not last eight days. He therefore "corrected" *octo dies* to *octoties*, a debased version of *octies*, that is, *eight times*, instead of *eight days*. Moreover, the event could not have happened in 807, and therefore Kepler corrected the date to 16 April *808*.[40] Maestlin took issue with this explanation but was unable to suggest a better one.[41]

Kepler calculated that at Mercury's inferior conjunction of 29 May 1607 the planet would be at its nodes, so that it should be visible as a black spot traversing the Sun. In order to witness this event, he began observing the Sun on 28 May by means of a camera obscura, in this case a dark attic where he held up a piece of paper to catch the Sun's rays admitted through a crack in the shingles. Indeed, a black spot was visible on the Sun's image. After checking the observation with others, Kepler concluded that he had observed Mercury on the Sun. He wrote this observation up in a little tract entitled *Phaenomenon Singulare seu Mercurius in Sole*, in which he showed a spot whose diameter was about one-twentieth the Sun's, and he tried to explain away this paradoxical *smallness*![42]

When in 1609 he finally managed to have *Astronomia Nova* as well as *Phaenomenon Singulare* printed, Kepler already had begun to depart markedly from the accepted distances, while he still agreed entirely with traditional notions about the sizes of heavenly bodies. He had the mathematical and physical ingredients that were to help make up his final, fully articulated system of sizes and distances. One new ingredient entered his thought in the following years, however: the evidence of the telescope. This new instrument would make him change his mind about planetary and stellar sizes.

* 7 *
Galileo and the Telescope

The history of astronomy can be divided into three overlapping epochs: naked-eye astronomy since before the dawn of civilization, telescopic astronomy beginning in 1609, and space astronomy beginning about 1960. In our story we have now reached the point at which the telescope was first used in astronomy. Although during the seventeenth century this new instrument radically changed Western ideas about the sizes and distances of celestial bodies, this change did not come suddenly or all at once, as we might expect; it took more than a generation. On the one hand, it took time for the telescope to become a routine part of the arsenal used by astronomers in their nightly assaults on the heavens; on the other hand, preconceived ideas about how large heavenly bodies ought to be often made astronomers indifferent to the evidence of this new instrument. Yet even in the first few years of its existence, in the hands of Galileo, the telescope helped cast doubt on received ideas about the sizes of heavenly bodies.

In October 1608 a small, eleven-page diplomatic newsletter was printed in The Hague. Its full title was *Ambassades du Roy de Siam envoyé à l'Excellence du Prince Maurice, arrivé à la Haye le 10. septemb. 1608*. Its title does not give any hint that this circular ends with an account of a spyglass on which a certain Hans Lipperhey, a simple maker of spectacles from the city of Middelburg in the province of Zeeland, had requested a patent that month. Besides stressing the obvious military value of the new device, the anonymous author of the tract also hinted at another use:[1]

> . . . the said glasses are very useful in sieges and similar occasions, for from a league and more away one can notice things as clearly as if they were quite near us; and even the stars that ordinarily do not appear to our sight and our eyes because of their smallness and the weakness of our sight can be seen by means of this instrument.

This report was distributed through Europe, and we know that Paolo Sarpi in Venice received it in November. The instrument itself followed quickly, and even before it reached many places the mere knowledge of its existence was enough to allow craftsmen and even scholars to reproduce it. By the summer of 1609, Thomas Harriot (1560–1621) near London was observing and mapping the Moon through a six-powered spyglass, and Galileo Galilei (1564–1642) in Padua was busily improving the device. By the end of that year Galileo had turned the feeble gadget into a powerful scientific instrument, and he was observing the Moon and stars with a twenty-powered telescope. In January 1610 he discovered four moons circling Jupiter. This epoch-making and entirely unexpected discovery became the centerpiece of *Sidereus Nuncius*, printed in March 1610, the first publication in the new field of telescopic astronomy.[2]

The main body of *Sidereus Nuncius* consists of four parts: a description of the new instrument, Galileo's discovery of the earthly nature of the Moon, his remarks on the appearances of the planets and fixed stars, and, finally and most important, an account of the newly discovered companions of Jupiter. We are concerned with the brief third part, for here Galileo began the assault on traditional ideas about the sizes of heavenly bodies. A ten-powered telescope magnified the Sun and Moon ten times, just as it did earthly objects, but when it came to the fixed stars and planets, things were different:[3]

> The stars, whether fixed or wandering, appear not to be enlarged by the telescope in the same proportion as that in which it magnifies other objects, and even the moon itself. In the stars this enlargement seems to be so much less that a telescope which is sufficiently powerful to magnify other objects a hundred-fold [in area] is scarcely able to enlarge the stars [and planets] four or five times [in area].

Up to 1609 or 1610, Galileo's own ideas about the sizes of fixed stars and planets had been entirely conventional.[4] He did not, however, reject the evidence of the new instrument. Rather, he gave an explanation of why fixed stars and planets seem to be magnified less than other objects by the telescope. In this explanation, for the first time in the history of astronomy, a distinction is made, albeit not a very clear one, between *size* and *brightness*.[5]

> When stars [and planets] are viewed by means of unaided natural vision, they present themselves to us not as of their simple (and, so to speak, their physical) size, but as irradiated by a certain fulgor and as fringed with sparkling rays, especially when the night is far advanced. From this they appear larger than they would if stripped of those adventitious hairs of light, for the angle at the eye is determined not by the primary body of the star [or planet] but by the brightness which extends so widely about it.

Now that the telescope had actually shown planets and stars stripped of their accidental rays at night, Galileo realized that observers had always been able to see them thus stripped without the aid of the new instrument, but only under suitable conditions:[6]

> . . . when stars [and planets] first emerge from twilight at sunset they look very small, even if they are of the first magnitude; Venus itself, when visible in broad daylight, is so small as

> scarcely to appear equal to a star of the sixth magnitude. Things fall out differently with other objects, and even with the moon itself; these, whether seen in daylight or the deepest night, appear always of the same bulk. Therefore the stars are seen crowned among shadows, while daylight is able to remove their headgear; and not daylight alone, but any thin cloud that interposes itself between a star and the eye of the observer. The same effect is produced by black veils or colored glasses, through the interposition of which obstacles the stars are abandoned by their surrounding brilliance. A telescope similarly accomplishes the same result. It removes from the stars [and planets] their adventitious and accidental rays, and then it enlarges their simple globes . . . so that they seem to be magnified in a lesser ratio than other objects. In fact a star of the fifth or sixth magnitude when seen through a telescope presents itself as one of the first magnitude.

This explanation was not very good. The telescope takes in more light than does the naked eye; it does not strip light away. Kepler sought the explanation not in the source but in the receptor, the eye. But his explanation was equally unsatisfactory.[7]

If the telescope did not treat stars and planets the same as other objects, there was also a distinction between stars and planets:[8]

> Deserving of notice also is the difference between the appearances of the planets and of the fixed stars. The planets show their globes perfectly round and definitely bounded, looking like little moons, spherical and flooded all over with light; the fixed stars are never seen to be bounded by a circular periphery, but have rather the aspect of blazes whose rays vibrate about them and scintillate a great deal. Viewed with a telescope they appear of a shape similar to that which they present to the naked eye, but sufficiently enlarged so that a star of the fifth or sixth magnitude seems to equal the Dog Star, largest of all the fixed stars.

In the cosmos of Aristotle and Ptolemy, there was no difference in kind between fixed stars and planets. The former were fixed while the latter wandered, but they were made up of the same substance and there was no systematic difference in their distances and motions. The fixed stars followed immediately above Saturn, and the speed of heavenly bodies decreased smoothly from a maximum for the Moon to near zero for the fixed stars. Copernicans had not been able to show any evidence to support their contention that the planets were like the Earth and that fixed stars were immensely far away. Galileo now carefully stressed the similarity between the planets and the Moon, which he had already shown to be Earth-like, and his evidence about the apparent sizes of the fixed stars strongly suggested that they were at an enormous distance from the Earth.

To avoid being scooped on such an earthshaking discovery as the moons of Jupiter, Galileo had rushed into print. *Sidereus Nuncius* was brief and qualitative. Galileo had shown how measurements could be made with the telescope,[9] but he reported no measurements of apparent diameters. Therefore, the important message that the traditional values for the apparent diameters of planets and fixed stars were much too large did not have much impact. What was needed was a complete set of telescopic measurements of the apparent diameters of the planets and large fixed stars. This would take time, for one had to wait for the planets to come into convenient positions for observations and measurements. Galileo did not make his discovery about the shape of Saturn until July 1610, and he verified the phases of Venus in the autumn of that year. The first survey of the planets had taken about a year. Just how far ahead of the competition Galileo was at this early stage is shown by the utterings of Kepler, who, after looking through a telescope made by Galileo, wrote:[10]

> . . . to us Jupiter, as Mars, and, in the morning, Mercury and Sirius, appeared four-cornered. For one of the angular diameters was blue-green, and the other purple, with in the middle a flaxen body with a wonderful fulgor. All this happens because of the weakness of sight, obscured in consequence of such dense light, as the instrument accumulates it.

Kepler's colleague at Heidelberg, Jacob Christmann (1554–1613) saw Jupiter ". . . completely on fire, so that it appeared separated into three or four fiery balls, from which thinner hairs spread in a downward direction."[11] Clearly, only Galileo and perhaps some of his associates were capable during these early years of supplying the world of science with a new set of apparent diameters, determined with the help of the telescope.

By the end of 1610 Galileo had finished a series of observations of Venus, culminating in his announcement that this planet waxes and wanes and goes through phases, just like the Moon. This meant that Venus goes around the Sun, so that sometimes it is "above" and sometimes "below" the Sun, a state of affairs that was predicted by the systems of Copernicus and Tycho but did not fit with the system of Ptolemy.[12] In his observations of Venus through a complete cycle of its phases, Galileo also noticed the dramatic variations in the planet's apparent diameter, and these variations, combined with his knowledge that the traditional, naked-eye estimates of planetary apparent diameters were much too large, apparently made him decide that an entirely new set of measurements of apparent diameters was needed. In February 1611 he wrote to Paolo Sarpi that he had looked into the method of determining the true measures of the planets and claimed that the traditional measures were in error by "more than 6,000 percent" in some cases.[13] Later we shall see how he arrived at this figure. Curiously enough, however, Galileo never produced a complete set of such measurements. Although for the rest of his life he vociferously took others to task for their faulty notions of planetary sizes, only twice (in print or correspondence) did he give an actual measurement of an angular planetary diameter.

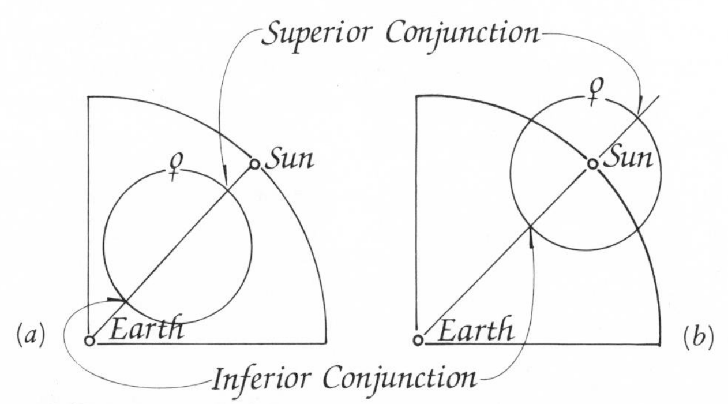

FIG. 11. Venus's position with respect to the Sun in the systems of Ptolemy (a) and Tycho Brahe (b)

The publication of *Tres Epistolae de Maculis Solaribus* in Augsburg, early in January 1612, signaled the start of a controversy between Galileo and the Jesuit Christoph Scheiner (1575–1650), who published his tract under the pseudonym of *Apelles latens post Tabulam*.[14] Galileo argued with Scheiner about the nature of sunspots and corrected him on the size of Venus.

As a follower of Tycho Brahe, Scheiner believed that Venus (and Mercury) goes around the Sun. Perhaps still unaware of Galileo's demonstration of Venus's phases, Scheiner tried to prove by another method that the planet is sometimes above and sometimes below the Sun. The ephemerides of Giovanni Antonio Magini predicted a superior conjunction of Venus with the Sun beginning on 11 December 1611 and lasting for 40 hours. At this particular conjunction the difference between the latitudes of the Sun and Venus was to be smaller than the angular diameter of the Sun, so Venus would pass in front of or behind the body of the Sun. As shown in figure 11a, in Ptolemy's scheme of the cosmos Venus is below the Sun both at inferior conjunction (when it is at the bottom of its epicycle) and at superior conjunction (when it is at the top of its epicycle). In the system of Tycho (and, for that matter, the system of Copernicus), on the other hand, Venus is below the Sun at inferior conjunction and above the Sun at superior conjunction (figure 11b). Before the telescope the body of Venus on the Sun had escaped astronomers, according to Scheiner, but the new instrument now made it possible to detect Venus on the Sun, especially since its apparent diameter was at least 3′. When he did not see Venus on the Sun during the superior conjunction, he concluded that the system of Ptolemy was incorrect.[15]

Now in all three world systems current at the turn of the seventeenth century, Venus's superior conjunction with the Sun occurred when Venus was at its greatest distance from the Earth. Scheiner had thus assigned Venus its traditional apparent diameter but at its greatest distance. Since, again regardless of the world system, the ratio of Venus's greatest to least distance from the Earth was 13:2 (see table 1, p. 27), this would mean that near inferior conjunction its diameter would approach 20′. Galileo did not miss the opportunity to correct Scheiner. In his first letter on sunspots, dated 4 May 1612, he wrote:[16]

> He supposes [Venus] ought to be seen [on the Sun] in the guise of a spot much larger than any [sunspot] we observe, saying that its visible diameter is three minutes of arc, and therefore its surface is one one-hundred-thirtieth that of the sun. With all due respect to him, this is not true. The visible diameter of Venus [at apogee] is not even the sixth part of one minute, and its surface is less than one forty-thousandth that of the sun, as in due course I shall make evident to anyone by direct experiment.

Scheiner had, in the meantime, revised his estimate of Venus's apparent diameter at superior conjunction to 1′ or at most 2′. This figure took the planet's greater distance into account, but it was still based on the pretelescopic figure. Still, he argued, Venus should have been seen on the Sun if a transit had occurred at superior conjunction, for innumerable sunspots whose diameters were hardly one-sixtieth, and sometimes as small as one-hundredth, of the Sun's were visible with the telescope.[17] Published in September 1612, this persistent "error" gave Galileo an opportunity to lecture Scheiner on the erroneous values of all traditional "measurements" of apparent diameters:[18]

> Against what we are shown by sense and experience, he vainly adduces the authority of men who were great enough in other respects, but who were quite mistaken about assigning Venus a diameter one tenth that of the sun. Partly they deserve to be excused, but not entirely. Their partial excuse is in the lack of the telescope, which has brought no small contribution to astronomy; but two things leave them open to criticism. One is that they ought to have observed the size of Venus by day and not by night, for its nocturnal headgear of rays makes it look ten times as large as in daytime when deprived of this. Thus they might easily have learned that the diameter of its tiny globe sometimes is not the hundredth part of the diameter of the sun. In the second place they should have distinguished one of its positions from the other, instead of indiscriminately pronouncing its diameter to be one tenth that of the sun. For when the planet is nearest the earth, its diameter is more than six times as great as when it is most distant, and although this difference can be precisely observed only with the telescope, it is nevertheless quite perceptible to the naked eye. In these regards, then, the astronomers cited by Apelles cease to afford him any support by their authority.

Thus, the lack of a telescope did not entirely excuse the "errors" of astronomers before 1609! It was unfair to expect them to have known that to the naked eye Venus's size is more accurately portrayed in daylight than in darkness. In his *Dialogue concerning the Two Chief World Systems* of 1632, Galileo would attempt to explain why they should have known this.

His second point, however, was well taken. In the *Planetary Hypotheses* Ptolemy had clearly stated that Venus's apparent diameter was one-tenth the Sun's, or 3′, *at Venus's mean distance*, that is, 622½ e.r. (see p. 25). Therefore, at perigee and apogee it should be 11¼′ and 1¾′ according to Ptolemy's own figures (see table 1, p. 27). Tycho's figures predicted much the same (see table 6, p. 50). A similar variation in size was predicted for Mars. Now, although to the naked eye Venus and Mars vary markedly in size/brightness, this variation does not appear to be as great as the ratio predicted by Ptolemy, and some medieval writers had already pointed this out (pp. 33, 39–40). Only Magini, late in the sixteenth century, had given apogee and perigee values for apparent diameters (see p. 52). The telescope had now dramatically revealed not only the phases of Venus, but also the exact variations in size predicted by the mathematical models. Although the theories of Tycho Brahe and Ptolemy, as well as Copernicus, all predicted the same variations in the sizes of Venus and Mars, Galileo argued that the observed variations were evidence for the Copernican system.[19]

Galileo repeated here that at superior conjunction or apogee Venus's apparent diameter was not even 1/200th of the Sun's.[20] Somehow he had ascertained that Venus's apparent diameter is 9″ at apogee and about 1′ at perigee. These are surprisingly accurate values, and they demonstrate that even with the primitive Galilean telescopes of 1610–12 (without micrometers, of course) a supremely talented observer such as Galileo could make fairly accurate measurements of angular diameters. These numbers also explain his statement to Sarpi, two years previously, that in some cases the pretelescopic values for apparent planetary diameters were in error by more than 6,000 percent (see p. 68). In his *Novae Coelestium Orbium Theoricae* of 1589, Magini had given Venus's apparent diameter at apogee as 9′ (see p. 52), that is, greater than Galileo had now determined it to be by a factor of 60, or 6,000 percent. With his flair for the dramatic, Galileo chose Magini as his straw man because of the excessive angular diameters his rival had assigned to the planets and fixed stars.

Scheiner still did not grasp the full implication of Galileo's argument, printed in *Istoria e Dimostrazioni intorno alle Macchie Solari* in 1613. In *Disquisitiones Mathematicae*, published in 1614, Scheiner allowed his student Johann Georg Locher to repeat Tycho's objection against the Copernican hypothesis (see p. 51) without discussing how the telescope affected the argument. Locher also gave Jupiter's apparent diameter as 2¾′,[21] a value taken straight from Tycho (see table 6, p. 50), despite the fact that Simon Marius or Mayr (1570–1624) earlier that year had published the first telescopic estimate of Jupiter's diameter, 1′,[22] and that

Scheiner himself was a frequent and careful observer of the heavens with the new instrument. Clearly, the telescope did not reveal at once the "errors" of the ancients when it came to the apparent diameters of the planets and fixed stars.

What was needed above all was an entirely new scheme of planetary diameters based on telescopic estimates or measurements. Curiously enough, however, such a scheme did not appear until 1633 (see chap. 8). Before that the telescopic measurements or estimates that appeared in print were: Galileo's measurement of Venus's apparent diameter in *Istoria e Dimostrazioni* of 1613; Marius's figure for Jupiter's apparent diameter in his *Mundus Iovialis* of 1614; and Kepler's statement in book 4 of his *Epitome Astronomiae Copernicanae* of 1621 that Jupiter's largest apparent diameter was 50″, Saturn's 30″, and Mars's "larger than Jupiter's, but not by much."[23] To this we may add the statement by Galileo in his unpublished "Letter to Ingoli" of 1624 that Jupiter's apparent diameter "scarcely attains 40″,"[24] another remarkably accurate figure. Estimates of the apparent diameters of fixed stars were even rarer.

Galileo addressed himself to Tycho's objection to the Copernican hypothesis for the first time in his "Letter to Ingoli," but he did not approach it head-on. Rather he assumed that a mediocre fixed star, say, one of the third magnitude, was equal in size to the Sun, so that its apparent diameter was simply a measure of its distance.[25] He elaborated on this argument in his *Dialogue* of 1632, singling out the version given by Scheiner and Locher in the *Disquisitiones Mathematicae*.

In the third day of the *Dialogue* Galileo reviewed all the telescopic evidence favoring the Copernican system. He repeated here the argument that the variations in the apparent diameters of Venus and Mars were evidence for the Copernican hypothesis, even though Kepler had taken him to task for this error in the appendix of his *Hyperaspistes* of 1625.[26] His reiteration of the smallness of the apparent sizes of the fixed stars and planets as they appeared through the telescope was to better effect. Many of his contemporaries seem to have been unaware of the full extent and importance of this discovery and its implications for Tycho Brahe's argument. According to Scheiner and Locher, in the Copernican scheme a fixed star would be larger than the *entire orbis magnus* of the Earth.[27] First, how long was the baseline of the measurement?[28]

> To begin with, I assume along with Copernicus and in agreement with his opponents that the radius of the earth's orbit, which is the distance from the sun to the earth, contains 1,208 of the earth's radii.

This number is not to be found anywhere in *De Revolutionibus*. As we have seen, it is the greatest solar distance calculated but not published by Erasmus Reinhold, subsequently recalculated from the *Prutenic Tables* by

Maestlin, and finally taken from Maestlin's appendix to Kepler's *Mysterium Cosmographicum* by Scheiner and Locher (see pp. 48, 60). But most important, this number is virtually the same as Ptolemy's solar distance in the *Almagest*. Galileo thus adhered to a very traditional absolute solar distance, and this has some strange implications.

Since Galileo did not quarrel with his predecessors on the distance of the Sun, his corrections of the apparent diameters of the planets meant equal corrections of their absolute diameters. Therefore, when he argued that the *apparent* sizes of planets were much smaller than had been thought up to that time, he was also arguing that their *absolute* sizes were much smaller. In fact, in his writing he did not distinguish between apparent and actual sizes. In his "Letter to Ingoli," he had argued that since Jupiter's diameter is only 40″, it is 30 times smaller than the Earth, not 80 times larger. Although, again, he had chosen Magini as his foil, even if he had used Tycho's apparent diameters, the Earth would have been one of the largest bodies in the solar system, roughly the same size as Jupiter and Saturn and much larger than the other planets.[29] Galileo's planetary sizes were as different from later measures, with which we are familiar, as they were from the traditional ones, so familiar to his predecessors and most of his contemporaries! His argument against Tycho's objection was as follows:[30]

> Secondly, I assume with the same concurrence and in accordance with the truth that the apparent diameter of the sun at its average distance is about one-half a degree, or 30 minutes; this is 1,800 seconds, or 108,000 third-order divisions. And since the apparent diameter of a fixed star of the first magnitude is no more than 5 seconds, or 300 thirds, and the diameter of one of the sixth magnitude measures 50 thirds (and here is the greatest error of Copernicus's adversaries), then the diameter of the sun contains the diameter of a fixed star of the sixth magnitude 2,160 times. Therefore if one assumes that a fixed star of the sixth magnitude is really equal to the sun and not larger, this amounts to saying that if the sun moved away until its diameter looked to be 1/2,160th of what it now appears to be, its distance would have to be 2,160 times what it is in fact now. This is the same as to say that the distance of a fixed star of the sixth magnitude is 2,160 radii of the earth's orbit. And since the distance from the earth to the sun is commonly granted to contain 1,208 radii of the earth, and the distance of the fixed star is, as we said, 2,160 radii of the orbit, then the radius of the earth in relation to that of its orbit is much greater than (almost double [that is, squared]) the radius of that orbit in relation to the stellar sphere. Therefore the difference in aspect of the fixed star caused by the diameter of the earth's orbit would be little more noticeable than that which is observed in the sun due to the radius of the earth.

In other words, at a distance 2,160 times as great as the radius of the Earth's orbit, a fixed star will show an annual parallax of $\tan^{-1} 1/1{,}080$ or 3′11″, which is only slightly more than the horizontal parallax of the Sun.

Tycho's point, however, had been precisely that such a quantity now lay within the discriminating power of astronomical instruments! If annual stellar parallax was as little as 1′, he had argued, his instruments would have revealed it. Since no stellar parallax could be detected, it had to be less than 1′, and therefore the distance of the fixed stars had to be greater than about 3,500 times the baseline of the measurement. For a Copernican that baseline was the radius of the Earth's orbit. Taking Galileo's values for the apparent diameters of fixed stars of the sixth magnitude, 5/6″, as well as his figure for the solar distance, 1,208 e.r., at the distance specified by Tycho a fixed star of the sixth magnitude would have a diameter of about 35 e.r., while our Sun had a diameter of 11 e.r. The Sun, therefore, would still be smaller than any of the fixed stars visible with the naked eye. It is difficult to tell how large a first magnitude star would be, for Galileo made conflicting statements about the distances and distribution of the fixed stars.[31]

If Galileo had not entirely disposed of Tycho's objection, he had at least reduced the problem to manageable proportions. The key lay in the apparent diameters of the planets and fixed stars as observed with the telescope. Here the ancients had been in error, and, as in his sunspot letters, Galileo held them fully responsible for their errors:[32]

> And truly I am quite surprised at the number of astronomers, and famous ones too, who have been quite mistaken in their determinations of the sizes of the fixed as well as the moving stars, only the two great luminaries being excepted. Among these men are al-Farghānī, al-Battānī, Thābit ibn Qurra, and more recently Tycho, Clavius, and all the predecessors of our Academician [i.e., Galileo]. For they did not take care of the adventitious irradiation which deceptively makes the stars look a hundred or more times as large [in area] as they are when seen without haloes. Nor can these men be excused for their carelessness; it was within their power to see the bare stars at their pleasure, for it suffices to look at them when they first appear in the evening, or just before they vanish at dawn.

The deception should have been especially obvious to them in the case of Venus when it is an evening star. For then it may be so small that it is difficult to see before sunset, while after sunset it shines with great brightness. How, then, could so many astronomers have been so wrong for so long?[33]

> To speak quite frankly, I thoroughly believe that none of them—not even Tycho himself . . . —ever set himself to determine and measure the apparent diameter of any star except the sun and moon. I think that arbitrarily and, so to speak, by rule

> of thumb some one among the most ancient astronomers stated that such-and-such was the case, and the later ones without any further experiment adhered to what this first one had declared. For if any of them had applied himself to making any test of the matter, he would doubtless have detected the error.

If we compare the various schemes of sizes and distances before Tycho and Magini, we see that although there was significant disagreement on the distances of heavenly bodies, there was none at all on their apparent diameters. These latter figures were canonical. Except for the last sentence, Galileo's conjecture about how this consensus came about is very astute, and it would, no doubt, have pleased him enormously to learn that these "errors" have now been traced back to Ptolemy and Hipparchus themselves. But how could the ancients have arrived at better figures for apparent diameters without the telescope?[34]

> It is true that the telescope, by showing the disc of the star bare and very many times enlarged, renders the operations much easier; but one could carry them on without it, though not with the same accuracy. I have done so, and this is the method I have used. I hung up a light rope in the direction of a star . . . and then by approaching and retreating from this cord placed between me and the star, I found the point where its width just hid the star from me. This done, I found the distance of my eye from the cord, which amounts to the same thing as one of the sides which includes the angle formed at my eye and extending over the breadth of the cord.

It was now a simple matter to measure the thickness of the cord and to calculate the angular diameter of the star. In this way Galileo had found the apparent diameter of a first magnitude star to be not more than $5''$, twenty-four times as small as Hipparchus and Ptolemy had made it. This method is vitiated by the fact that the diameter of the pupil is of roughly the same size as the diameter of the cord. Galileo tried to deal with this problem, insisting that his procedure produced reliable results.[35]

Convincing as his argument in the *Dialogue* was, Galileo did not go on to supply the scientific community with a set of telescopically measured apparent diameters. This is baffling, for the two planetary measures he did give, Venus and Jupiter, were tantalizingly good; better than his successors could do for some time. Furthermore, on the issue of planetary distances, a subject that clearly did not interest him very much, his ideas seem to have been entirely in keeping with tradition. The result was that the solar system implied in his arguments, although never made explicit, was one in which the distances were much the same as Copernicus's but, except for the Sun, Moon, and Earth, the planets had shrunk dramatically. Because he did not make the appropriate measurements, his arguments remained qualitative; it would take more than qualitative arguments to discredit the traditional, and still canonical, scheme of sizes and distances.

The great Florentine's failure to produce a new system of cosmic dimensions should not blind us to his great contribution to this subject. If his consistent message that the traditional sizes of the planets and fixed stars were much too large did not have the immediate impact we might have expected, it is largely because Galileo was very far ahead of the competition when it came to telescopic matters. Not only did his colleagues labor with inferior instruments, they also fell far short of Galileo's uncommon talent for observing. Thus, in 1610–11 he had already been able to determine Venus's apparent diameter with a degree of accuracy that was not matched for a generation, while Christoph Scheiner, whose sunspot observations in *Rosa Ursina* (1630) show him to be a rather good observer, never managed to get a measure of Venus's apparent diameter that was independent of the traditional value.

When good research telescopes became somewhat more common in the 1630s and astronomers began to use these instruments routinely, rather than looking through them only on special occasions, the impact of Galileo's message was felt. It was only a matter of time before the naked-eye estimates of planetary and stellar apparent diameters made by Hipparchus and Ptolemy—and accepted by their successors for many centuries—were utterly discredited, first in astronomical circles, and then in wider circles of the educated world. In the meantime, one completely new system of cosmic dimensions that took telescopic observations into account had already appeared in the literature. This system was one of Kepler's great achievements.

* 8 *
Kepler's Synthesis

Astronomy was anything but a static discipline when the telescope came into use. Whereas in 1500 there had been only the Ptolemaic system of astronomy, by 1600 the systems of Copernicus and Tycho provided increasingly viable alternatives. At the turn of the seventeenth century the science of astronomy was in the middle of a revolution, and the telescope changed the terms of the debate and the course of the revolution. This change is most obvious in the new information, such as the satellites of Jupiter, brought into astronomy by the new instrument. It is also clearly evident, however, in the theoretical work of Kepler.

As discussed in chapter 6, Kepler had begun his quest for a new causal and mathematical explanation of the cosmos in the mid-1590s, and he had published his first approximation of a new cosmology in his *Mysterium Cosmographicum*. He continued his quest in the new century, aided now by the accurate observations of Tycho Brahe, and eventually, in his *Epitome Astronomiae Copernicanae* (1618–21), he published a mature system of the world based on elliptical astronomy, celestial physics, and mathematical harmonies. In this Keplerian synthesis, telescopic evidence played a significant role.

When Kepler read *Sidereus Nuncius* in April 1610, he realized that the new instrument with which Galileo had made his discoveries would change astronomy profoundly. In his *Dissertatio cum Nuncio Sidereo,* sent to Galileo on 19 April and printed the following month, Kepler gave his initial enthusiastic reaction to the new discoveries. The telescope, he felt, could be very useful in the problem of sizes and distances:[1]

> I want your instrument for the study of lunar eclipses, in the hope that it may furnish the most extraordinary aid in improving, and where necessary in recasting, the whole of my "Hipparchus" or the demonstration of the sizes and distances of the three bodies, sun, moon, and earth. For the variations in the solar and lunar diameters, and the portion of the moon that is eclipsed, will be measured with precision only by the man who is equipped with your telescope and acquires skill in observing.

Yet, it would be a mistake to think that Kepler immediately understood what the telescope meant for the apparent diameters of planets and fixed stars. As is apparent in the *Dissertatio*, he still believed that the majority of the fixed stars in the catalogs were larger than 1′.[2] The smallness of planets and fixed stars, as well as the distinction between the two, as revealed by the telescope could, it appears, only be appreciated by a good observer who had scrutinized these bodies through a telescope as powerful and as optically excellent as Galileo's. By his own admission, Kepler was a poor observer, and his reactions to *Sidereus Nuncius* were based at best on observations with common spyglasses made by makers of spectacles.[3] As we have seen in chapter 7, even when he was able to observe with a telescope made by Galileo later that year, he was unable to verify Galileo's statements about the sizes of the planets and fixed stars. It is not surprising, therefore, that Kepler's mature thoughts about the sizes of the planets and fixed stars, although they took account of what little telescopic information was available, were based mainly on physical and harmonic notions.

In establishing a scheme of absolute sizes and distances in the systems of Copernicus or Tycho, the most difficult problem was the determination of one absolute distance. When he wrote his *Astronomia Nova*, Kepler was convinced that Mars's parallax at opposition was smaller than the margin of error of Tycho's measurements, and this meant that solar parallax had to be less than the traditional value of 3′. But how much less? From his analyses of lunar eclipses he could only deduce that the solar distance had to be greater than 700 e.r. and less than 2,000 e.r., and he clung to these limits for the next decade, using a figure of 1,432 e.r.—corresponding to a solar parallax of 2′24″—in *De Stella Nova* of 1606 (see pp. 61–62).

By 1615 Kepler's faith in the eclipse diagram, the traditional method for determining the solar distance and parallax, had eroded. One cannot do predictive astronomy, however, without using some figure for solar parallax, and when Kepler began his work on the ephemerides he had to arrive at a value. In the introduction to the ephemerides for 1617, he explained how he had come to reject the old figure:[4]

> I have fixed the horizontal parallax of the Sun two-thirds as large as the Tychonic one, because Tycho himself did not demonstrate his [parallax] but assumed it to be about as large as it follows from the Ptolemaic demonstrations, except that he changed it a very little, so that the noble number 576, cajoled from the Pythagorean philosophy, would obtain in the interval between the Sun and Earth. But I have been forced, partly by the contemplation of Mars' parallaxes, as can be seen in the Commentaries of Mars [i.e., the *Astronomia Nova*], and partly by the method based on solar and lunar eclipses, to reduce the parallaxes of the Sun.

This was not to say that Kepler had arrived at his new value for the solar distance by the old method. His study of the eclipse diagram had shown him that this method was worthless. Ptolemy had been wrong in setting the Sun's constant apparent diameter equal to the smallest apparent diameter of the Moon, thus ruling out annular solar eclipses, while his estimates of the extent to which the Moon was eclipsed during various eclipses had been nothing more than choices of convenient values:[5]

> Nor is the force of the Ptolemaic demonstration so great that the quantity of these parallaxes of the Sun produced by him would have to be treated scrupulously, since he involves an impossible [thing] about the diameter of the Sun and also assumes, or rather, fits the estimate of the digits of eclipses of the Moon, an uncertain business which hardly reaches those small [quantities.]

Having nothing more accurate to put in the place of the eclipse diagram, Kepler resorted to mathematical harmony to pick a solar distance of about 1,800 e.r., which he made correspond to a horizontal solar parallax of 2′:[6]

> And therefore, according to me, the Sun ascends to a height of 1,800 semidiameters of the Earth; according to Tycho less than 1,200. Moreover, the round number 1,800 pleased me as it is about thirty-fold the distance of the Moon from the Earth, which is nearly 60 [e.r.], and thus the proportion of the intervals between Sun and Earth and between Moon and Earth is the same as that of the revolutions of the Earth and Moon about the same center of the Earth.

A few pages later in the same ephemerides Kepler wrote that in making up his lunar equations he had chosen a ratio of solar to lunar distances of 23:1, which corresponds to a solar distance of about 1,380 e.r. The result was that the Moon's face was exactly bisected 2½° before quadrature. He only wished to use this ratio for this particular purpose, and he felt that even if it introduced an error of half a degree in the location of the Moon's various phases, such an error would not be important.[7] Obviously, Kepler was now toying with an even more venerable method of finding solar distance and parallax, Aristarchus's lunar dichotomy.

With the better lunar theories of Copernicus and especially Tycho, the Moon's absolute distances were known more accurately than before, and all one needed was the ratio of solar to lunar distance to arrive at the absolute distance of the Sun. That the telescope might be useful in refurbishing the method of lunar dichotomy to arrive at this ratio had not been lost on some early telescopic observers. Thus, on 21 January and again on 19 April 1611, Thomas Harriot, who had corresponded with Kepler from 1606 to 1609, had tried to determine the exact moment of dichotomy and the Moon's distance from quadrature at that time. His trials, however, had shown that the new instrument did not make this ancient method more accurate.[8] But Kepler did not know about Harriot's efforts.

In his ephemerides for 1618—written after those for 1617 but printed six months earlier in October 1617—Kepler took this matter up in earnest. He mentioned that he had retained the 23:1 ratio between the solar and lunar distances, but this ratio and his solar parallax were still uncertain. He had again used a horizontal solar parallax of 2′, but he was now convinced from considerations of harmony that it was in reality no greater than 1′. The interval between lunar dichotomy and quadrature would, therefore, be smaller than the time obtained from these ephemerides. For a measurement of the exact length of that interval he appealed to better equipped observers:[9]

> . . . I leave [this interval] to be determined by Galileo and Marius with [their] eyes and the telescope, if you will trust it in this great matter, especially because of the unevenness of the lunar globe. Take pains, men of philosophy! The matter is not to be disdained and entirely worthy . . . , that is, from a mere look at the exactly bisected Moon and an annotation of the time, accurate to the hour, and from quadrature of her, to pronounce on the proportion of the orb of the Moon and even on the quantity of the parallaxes of the Sun, and, finally, to pronounce on the proportion itself of the altitudes as well as the globes of all the planets to the semidiameter of the Earth. All of which until now depended on the treacherous shadow of the Earth and on an estimate of disgusting diversity of the digits removed from the Moon, and finally on a confusion of the many causes of the ingredients of the demonstration, which single [factors] can be separately corrected, with the rest remaining fixed, with an incredible variety of parallaxes of the Sun resulting.

By 1617 Kepler had completely abandoned the eclipse diagram for determining the Sun's distance.[10] The only method he still considered feasible for deriving solar distance and parallax directly from the phenomena was that of lunar dichotomy. A solar parallax of 2′, such as he used in these ephemerides, meant that quadrature would precede or follow dichotomy by about 4 hours; a solar parallax of 1′ would decrease these intervals to about 2 hours.[11]

Aristarchus's method, long thought to be unreliable, had thus been resurrected by Kepler because he thought that the telescope would allow observers to establish the exact moment of dichotomy accurately. It was thought that, knowing what feature lies precisely on the middle line of the disk, an observer could easily determine the moment when the terminator passed over that feature. Kepler himself tried this method, but he was disappointed by the results. To jump ahead of our story, in the second part of the ephemerides, comprising the years 1621 to 1626 but not published until 1630, Kepler wrote:[12]

> But the observation or apprehension of this phase through the telescope, on which I advised in the preface to the ephemerides for the year 1619, I find very difficult and entirely confused be-

TABLE 10 Heliocentric Distances of the Planets Based on Tycho's Observations and Elliptical Orbits

Planet	Distance from the Sun in a.u.		
	Least	Mean	Greatest
Mercury	0.307	0.389	0.470
Venus	0.719	0.724	0.729
Earth	0.982	1.000	1.018
Mars	1.382	1.523	1.665
Jupiter	4.949	5.200	5.451
Saturn	8.968	9.510	10.052

> cause of the unevenness of the Moon, so that at the same moment one part of the division [terminator] appears concave and the rest gibbous. Indeed, this interval of time so small can afford us no aid in the determination of the exact quadrature and through it of the true proportion of the orbs of the Sun and Moon.

At best, Kepler thought at the end of his life, telescopic observations of the Moon could perhaps show that the 6-hour interval between dichotomy and quadrature predicted by Aristarchus and Ptolemy was wrong.[13] For actually determining the correct ratio of solar to lunar distances, however, Aristarchus's method was useless. In fact, no method could produce solar distance and parallax from the phenomena with acceptable accuracy and uniformity. This is not to say that Kepler had given up on finding the measure of the universe—far from it.

* * *

In 1618 Kepler turned from his work on the tables to the contemplation of harmony once again and produced his *Harmonice Mundi*, published the following year. Fundamental to this study were accurate figures for the planetary distances measured from the true Sun. By 1618, after nearly 17 years of studying the very accurate observations made by Tycho, Kepler had arrived at the planetary distances given in table 10, more exact than any previous distances.[14]

Using these distances, he found on 15 May 1618 that "the ratio between the periodic times of any two planets is precisely the sesquialter of the ratio of their mean distances,"[15] or, in modern terms, the squares of the periods of the planets are proportional to the cubes of their heliocentric distances. This is Kepler's third law. The periods of the planets had always been subject to rather precise determination; the Copernican distances had been more difficult. If we test his third law on the figures given in his *Harmonice Mundi* (see table 11), we see that Kepler's fit was accurate to three significant figures, except in the case of Saturn, and even in this case the error was less than 1 percent.

TABLE 11 Kepler's Third Law Illustrated from the Planetary Distances and Periods in *Harmonice Mundi*

Planet	R Mean Heliocentric Distance (a.u.)	T Period (yrs)	R^3	T^2
Mercury	0.389	0.241	0.058	0.058
Venus	0.724	0.615	0.379	0.378
Earth	1.000	1.000	1.000	1.000
Mars	1.523	1.881	3.53	3.53
Jupiter	5.200	11.862	140.6	140.7
Saturn	9.510	29.457	860.1	867.7

The third law was a remarkable achievement, but by itself, as a mere correlation, it could mean little to Kepler. There had to be harmonic (that is, mathematical) and physical reasons for the distances and the relationship between them. He laid out the harmonic rationale in the *Harmonice Mundi*, and he gave a complete mathematical-physical account of the sizes and distances of all bodies in the universe in the fourth book of his *Epitome Astronomiae Copernicanae*, written immediately after *Harmonice Mundi* and published in 1620.

Although the five regular solids were still very much in evidence because they gave a reason for the number of planets and a rough estimate of their distances from the Sun, they did not produce distances that were sufficiently accurate. Kepler now began searching for harmonious relationships between both the aphelion and perihelion distances of successive planets. With the distances themselves he had little luck, but the daily angular motions at aphelion and perihelion as seen from the Sun produced ratios that were reasonably harmonious. After making refinements and adjustments that drew him heavily into state-of-the-art musical theory, Kepler produced planetary velocities at aphelion and perihelion entirely from harmony. From these velocities he then calculated eccentricities and finally greatest and least distances from the Sun. The results were impressive (see table 12).[16]

This was as far as Kepler went in the *Harmonice Mundi*, but in the fourth book of the *Epitome* he presented his complete system of sizes and distances of the heavenly bodies. This scheme was based on mathematical and physical principles, or, as Kepler himself expressed it:[17]

> Herein the natural and archetypal causes of *Celestial Physics* that is, of all the magnitudes, movements, and proportions in the heavens are explained and thus the Principles of the Doctrine of the Schemata are demonstrated. . . . This book is designed to serve as a supplement to Aristotle's *On the Heavens*.

Kepler divided the planets into primary and secondary ones. There were six primary planets and seven secondary ones, our Moon, Jupiter's

TABLE 12 Planetary Distances Derived from Musical Harmonies Compared to Those Calculated from Tycho's Observations (*Harmonice Mundi*)

	Heliocentric Distance in a.u.			
	from Harmonies		from Observations	
Planet	Greatest	Least	Greatest	Least
Saturn	10.118	8.994	10.052	8.968
Jupiter	5.464	4.948	5.451	4.949
Mars	1.661	1.384	1.665	1.382
Earth	1.017	0.983	1.018	0.982
Venus	0.726	0.716	0.729	0.719
Mercury	0.476	0.308	0.470	0.307

four moons, and the two lateral bodies that accompanied Saturn. Of these secondary planets only the Moon revolves around the Earth, our home, and therefore it was the only secondary planet that had to be incorporated in the proportions; the others could be ignored.[18]

Since the Earth is the home of a speculative creature, created in God's image, and since in the beginning God created the heavens and the Earth, the inquiry should start from the Earth. Yet, the Sun is the first of all created bodies, if not in the temporal order, then in the archetypal order, for it is the measure of all things:[19]

> . . . the first body, because first, does not acquire any ratio to the bodies following; but rather the bodies following acquire a ratio to it as first. Wherefore there is no archetypal cause of the magnitude of the sun: nor could there have been a different globe twice as great as it is now, because the rest of the universe and the whole world and man in it would have had to be twice as great as they are now.

The magnitude of the Sun was, therefore, the first that had to be determined. Its apparent diameter seen from the Earth is the 720th part of the circle of the zodiac, a number divisible into many harmonious parts. From this it follows that the distance from the Earth's center to the Sun's is 229 times the solar radius, for tan $15' = 1/229$.[20] This was not enough, however, to determine the proportion of the sizes of Sun and Earth. Kepler now stated a harmonic requirement:[21]

> Doubtless, because the Earth was to be the home of a measuring creature, the same Earth became the measure, by its body of the bodies of the world, and by its semidiameter as a line, of their lines, that is, their intervals. But since the mensuration of bodies is distinct from the mensuration of lines, and since the proportion between the bodies of the Earth and Sun is first and the proportion between the diameter of the Earth and the interval between the Earth and the Sun is also first, nothing is more in accord with the correct, elegant, and ordained order [*contemporatio*] than that the equality of the two proportions is es-

> tablished, so that as many times as the body of the Earth is contained in the body of the Sun, so many times also the Earth's semidiameter is contained in the interval between the centers of the Sun and the Earth: just as the Earth's body is to the Sun's body, so the Earth's semidiameter is to the distance between the centers.

If the ratio of the volumes of Sun and Earth is as the ratio of the solar distance to the Earth's radius, then we can write:

$$R^3{:}r^3 = D{:}r,$$

where R is the Sun's radius, r the Earth's radius, and D the solar distance. Since r is the unit in terms of which we wish to express R and D,

$$R^3 = D.$$

But we also know from the Sun's angular radius that

$$D = 229\,R.$$

So that $R = 15$ e.r. and $D = 3{,}469\frac{1}{3}$ e.r. But this solar distance corresponds to a horizontal solar parallax of 59.45″, or 1′. Here, then, was the harmonic reason for a solar parallax of 1′ alluded to in the ephemerides of 1618 (see p. 80).[22]

In the case of the Moon, made to obscure the Sun, a similar relationship applied:

$$r^3{:}x^3 = d{:}r,$$

where x is the Moon's radius and d the lunar distance. Expressed in Earth radii,

$$x^3 = \frac{1}{d}.$$

Since the Moon's angular diameter is the same as the Sun's,

$$d = 229\,x,$$

and thus, $x^4 = \frac{1}{229}$, and $x = 0.257$ e.r. From this, the lunar distance, d, is slightly less than 59 e.r. Hence the lunar distance was the mean proportional between the Earth's radius and the solar distance (1:59 = 59:3,469), and the Sun's radius was to the Moon's as the Moon's distance was to the Earth's radius.[23] While the solar parallax thus derived confirmed the figure Kepler had arrived at by other means, the lunar distance here derived was the same as the lunar apogee distance at quadrature found by Tycho.[24]

The sizes of the planets presented Kepler with a problem. He felt that "nothing is more in concord with nature than that the order of the sizes should be the same as the order of the spheres."[25] In other words, their sizes should be proportional to their distances from the Sun. But how were the planetary sizes to be expressed? By diameters, surface areas, or volumes? In the *Harmonice Mundi* Kepler had stated:[26]

> . . . it is a very probable conjecture (as relying on geometrical demonstration and the doctrine concerning the causes of the planetary movements given in the *Commentaries on Mars*) that

> the volumes [*moles*] of the planetary bodies are in the ratio of the periodic times.

This meant that the volumes were proportional to the 3/2 power of the distance, so that the surface areas were proportional to the distances themselves. When his friend and correspondent Johannes Remus Quietanus visited Kepler in 1618, however, he brought with him telescopic estimates of the apparent planetary diameters, and these figures did not support Kepler's a priori approach based on mathematical harmony and Aristotelian physics. Kepler was too good an astronomer to ignore Remus's estimates, although he was under no illusions about the accuracy of these figures:[27]

> Of these three modes [that is, diameters, surface areas, or volumes], the first is refuted beyond controversy both by the archetypal reasons and also by the observations of the diameters made with the help of the Dutch telescope; up to now I have approved of the second mode, and Remus Quietanus of the third. On my side the better reasons, the archetypal, seemed to stand; on Remus' side the observations stand; but in such a delicate question I was afraid that the observations were not certain enough not to be taken exception to.

Each of the three modes predicted a set of apparent planetary diameters at opposition as seen from the Earth. For the superior planets the prediction is as follows, in rough numbers, assuming Kepler's solar distance of 3,469 e.r.:

	1st mode	2d mode	3d mode
Mars	7½′	6⅛′	5¾′
Jupiter	2⅔′	1⅐′	6/7′
Saturn	2½′	⅚′	9/16′

Although Kepler did not give these explicit predictions, these were the numbers to be tested against Remus's telescopic estimates. Kepler wrote:[28]

> But I yielded to Remus and the observations. For Jupiter at opposition and at the perigee of its eccentric often appeared to me to occupy about 50″; Remus estimates Saturn to occupy 30″; and Mars, at opposition and perigee in Aquarius, appears larger than Jupiter, but not by much.

Obviously the proportionality of distances and volumes gave the best results. In the cases of Jupiter and Saturn the fit was virtually perfect, while in the case of Mars agreement was poor in all three modes, but better, at any rate, in the third.

Moreover, the proportionality of distances and volumes of the planets was in harmony with the previously determined proportions of the Sun, Earth, and Moon:[29]

> . . . in the same way that we have already assumed that the ratios between the bodies of the Sun and the Earth, [and] of the

> Earth and the Moon were the same as [the ratios between] the semi-diameter of the Earth and the semi-diameters of the spheres, so, now, we must assume that the ratios between the planetary bodies follow those of the semi-diameters of the spheres. Thus, Saturn by the volume [*moles*] of its body will be a little more than ten times larger than the Earth, Jupiter more than five times, and Mars one and one-half times; but [the volume of] Venus will be a little smaller than three-quarters of that of the Earth, that of Mercury a little larger than one-third.

If the sizes and distances of the planets had now been established, Kepler was by no means finished. He proceeded to give a physical, that is dynamic, rationale of this scheme which could also produce his (third) law relating the distances of the planets to their periods. In his *Mysterium Cosmographicum* he had introduced an *anima motrix* emanating from the rotating Sun and pushing the planets along their paths, and in the *Astronomia Nova* he had specified that this force emanates not in a sphere but rather in a plane, the plane of the solar system.[30] Kepler now argued that the amount of *species* of this motive power intercepted by a planet at any time depended on the planet's volume:[31]

> For as is the case in a mill, where the wheel is turned by the force of the stream, so that, the wider and longer the wings, planks, or oars which you fasten to the wheel, the greater the force of the stream pouring through the width and depth which you will divert into the machine; so too that is the case in this celestial vortex of the solar *species* moving rapidly in a gyro—and this form causes the movement. Consequently the more space the body . . . occupies, the more widely and deeply it occupies the motor virtue . . . and the more swiftly, other things being equal, is it borne forward; and the more quickly does it complete its periodic journey.

Since, as we have seen above, the volumes vary with the distances, while the strength of the *anima motrix* varies inversely with the distance, these two effects exactly cancel each other, and all planets are pushed along by equal forces. This leaves only the lengths of the paths and the weights of the planets to account for the differences in periods.[32]

The paths vary directly with the distances from the Sun, so if all planets had the same quantity of matter their periods would vary directly with their distances. In fact, however, the periods vary as the $3/2$ power of the distances, and Kepler had to pick appropriate densities to account for this. If we let T be the period of a planet, C the circumference of its orbit, r its distance from the Sun, v its velocity, F the force by which it is pushed along its path, q its density, V its volume, and W its weight, then

$T \propto r^{3/2}$, $C \propto r$, and $v = C/T$,
therefore $v \propto r^{-1/2}$.

From Aristotle we know that $v \propto F/W$, but since the forces pushing each planet along are exactly equal,

$v \propto 1/W$, and thus $W \propto r^{1/2}$.

But $W = qV$ while $V \propto r$, so that:

$q \propto r^{-1/2}$.

In other words, the densities of the planets must vary inversely as the square root of their distances.[33] Accordingly, Kepler assigned the following relative densities:[34]

Saturn	324	the hardest precious stones
Jupiter	438	the lodestone
Mars	810	iron
Earth	1,000	silver
Venus	1,175	lead
Mercury	1,605	quicksilver
Sun	1,800 or 1,900	gold

The entire system of planetary sizes and distances was thus complete. It combined physics with harmonious proportions and telescopic observations (at least for the superior planets). And if the telescopic observations of Mars did not fit the scheme very well, it was the only snag in an otherwise entirely coherent scheme.

It remained for Kepler to give an account of the fixed stars. He again used the proportion set out in *De Stella Nova* of 1606: just as the moved is the mean proportional between the mover and the place, so the radius of the sphere of Saturn is the mean proportional between the radius of the Sun and the sphere of the fixed stars.[35] From the Sun's apparent size it follows that its radius is contained 229 times in the radius of the Earth's sphere ($\sin 15' = 1/229$). Since Saturn is almost 10 times as far from the Sun as the Earth is, we can set the radius of its sphere equal to about 2,000 solar radii. This will make the radius of the sphere of the fixed stars $2{,}000^2$, or 4,000,000 solar radii, or 60,000,000 e.r., since the Sun's radius is about 15 e.r.[36]

If the reader found such a distance hard to accept, he had to agree, at least, with Tycho's measurements of the position of the polestar, which showed that if there was an annual stellar parallax it was imperceptible, that is, less than 1′. This meant that, according to Tycho's own calculations, in the Copernican system the distance of the fixed stars contained at least 3,500 *diameters* of the Earth's sphere ($\sin 1' = 1/3500$). Given Kepler's solar distance of 3,469 e.r., this worked out to be over 24,000,000 e.r.[37] The measurements could not be more accurate than 1′, "since the diameter of the polestar seems to be equal to at least 1′,"[38] an assessment with which Galileo would surely have disagreed. Kepler went to some pains to point out just how very small Tycho's universe was compared to those of his predecessors. If one recalculated the distance of the fixed stars according to the nesting sphere principle, while taking into account the diameters of the planets, then even Ptolemy made the cosmos twice as large as Tycho, 25,742 e.r.[39]

In the preface to his *Mysterium Cosmographicum*, Kepler had mentioned ". . . the splendid harmony of those things which are at rest, the Sun, the fixed stars and the intermediate space, with God the Father, and the Son, and the Holy Spirit."[40] He now quantified this notion by supposing that these three parts of the cosmos each contained the same quantity of matter. Their densities were different, however: that of the Sun might be compared to gold, that of the intermediate space with air, and that of the sphere of the fixed stars with the density of an aqueous, vitreous, or crystalline substance. At any rate, the density of the matter of the sphere of the fixed stars was the mean proportional between the densities of the Sun and the ethereal region.[41]

Since the radius of the sphere of the fixed stars contains 4,000,000 solar radii, the volume of the space contained in it, that is, the volume of the ethereal region, is to the volume of the Sun as $4{,}000{,}000^3$ is to 1, or, using modern notation, $64 \times 10^{18}:1$. The volume of the shell in which the fixed stars are contained is the mean proportional between these two figures, or 8×10^9 solar volumes. In order to find the thickness of the sphere or shell of the fixed stars, Kepler first added the volumes of the three regions together:[42]

ethereal region	64×10^{18}
sphere of the fixed stars	8×10^9
Sun	1
Volume of the cosmos	$64 \times 10^{18} + 8 \times 10^9 + 1$

This, then, was how many times the solar volume was contained in the cosmos. Therefore, the cube root of this figure will be the number of times the solar radius is contained in the radius of the universe. Kepler made the cube root 4,000,000$\frac{1}{6000}$,[43] so that the radius of the universe was 4,000,000$\frac{1}{6000}$ solar radii. He had already determined the inner radius of the sphere of the fixed stars to be 4,000,000 solar radii. This left only $\frac{1}{6000}$ solar radii for the thickness, or rather thinness, of the spherical shell containing the fixed stars! Knowing that the solar radius contained about 15 e.r., and that the radius of the Earth was 860 German miles according to his teacher Maestlin,[44] Kepler calculated the thickness of this shell to be about 2 German miles, which corresponds to about 9 English miles.[45]

This would mean, however, that the fixed stars were incredibly small in Kepler's universe. Returning once again to his refutation of Bruno (see p. 63), Kepler argued that if one observed the Sun from one of the fixed stars, it would subtend an angular diameter of $\frac{3469}{60{,}000{,}000} \times 30'$, or $\frac{1}{600}'$, and yet from that enormous distance it would appear brighter than any fixed star appears to us:[46]

> For if, for example, you pierce through a wall with only a pin, so that the sun can shine through the hole, a greater brightness

> is poured through from the beams than all the fixed stars shining together in a cloudless sky would give. And the eye is not injured by any of the fixed stars; but it cannot bear to look towards the sun even from a distance.

The Sun was, therefore, much larger than any of the fixed stars, and the diminutive nature of the fixed stars was further supported by telescopic observations:[47]

> Skilled observers deny that any magnitude as it were of a round body can be uncovered by looking through a telescope; or rather, if a more perfect instrument is used, the fixed stars can be represented as mere points, from which shining rays, like hairs go forth and are spread out.

In fact, even the best telescopes of Kepler's day hardly showed the fixed stars as very sharply defined round dots, let alone dimensionless points of light. The rays and fringes surrounding their images hid any sharp outline. Kepler was led to this statement by his archetypal argument which led to incredibly small actual sizes of the fixed stars. It took several generations of telescopic observations before most astronomers were convinced that, seen from the Earth, the fixed stars appear as dimensionless points.

In the meantime, however, Kepler's conclusion was a neat refutation of Tycho's objection to the Copernican system. He had shown that the fixed stars are not many times greater than the Sun, but are in fact many times smaller. In his scheme of sizes and distances he at once dramatically expanded the universe and refuted Tycho's argument.

Kepler's enlargement of the universe was the combined result of his solar distance and Tycho's measurements of the positions of the polestar. Since these measurements showed no annual stellar parallax, and since they were accurate to 1′, the *lower limit* of the distance to the fixed stars was 3,500 times the diameter of the Earth's orbit. A Copernican who accepted one of the traditional solar distances would still have to make the distance to the fixed stars at least about 8,000,000 e.r.

The solar distance of 3,469 e.r. resulted in a harmonious system of sizes and distances, but it was also based on a more practical aspect of astronomy: a solar parallax of 1′ or less gave better results in the calculations of eclipses. We must, therefore, not be misled by the great increase in the solar distance instituted by Kepler—a radical break with tradition—into thinking that it was his aim to make the solar distance, and therefore the stellar distance, as large as possible. The exact opposite is true: a solar distance of 3,469 e.r. was the *minimum* distance Kepler could accept because a solar parallax of 1′ was the *greatest* parallax he could subscribe to. His analysis of the parallax of Mars, and his calculations of eclipses which he could compare with the observations, had convinced him that a solar parallax of more than 1′ gave unsatisfactory results. A figure of 1′ or less would leave the disagreement between calculation and observation within the margin of error. Nothing, then, except perhaps an attachment to the traditional scale of things, stood in the way of a further decrease of the solar parallax; a value of 30″ or even 15″ gave equally good results. Kepler's archetypal and harmonic arguments anchored the solar parallax at the largest value he could defend!

* * *

Before discussing Kepler's ideas on solar distance and parallax further, we must introduce another factor which became central to positional astronomy in the seventeenth century: the problem of atmospheric refraction. Solar parallax is a correction, varying from a maximum at the horizon (the horizontal solar parallax) to zero at the zenith, which must be made to the Sun's observed altitude above the horizon in order to obtain its true, geocentric altitude. This is not, however, the only correction that has to be made. As light rays from celestial objects enter the Earth's atmosphere, they are bent toward the normal. Therefore the effect of refraction is to *raise* the apparent position of a celestial body while the effect of parallax is to *lower* it. Refraction, too, varies from a maximum at the horizon to zero at the zenith. Its maximum is slightly more than half a degree, an order of magnitude larger than the horizontal parallaxes of all heavenly bodies except the Moon. It is ironic that although Ptolemy was perfectly familiar with the phenomenon of refraction as light leaves one medium and enters another, he did not correct measured positions for its effect.[48]

Until the sixteenth century astronomers followed Ptolemy's example, although they, too, knew about the effects of refraction. They dutifully made corrections for parallax in the Sun's observed positions while ignoring what turned out to be the much larger effect of refraction. It was Bernard Walther (1430–1504), the student of Regiomontanus in Nüremberg, who first drew attention to the importance of refraction in solar positions, and it was Tycho Brahe who instituted the practice of correcting his observed positions for this effect.[49]

Tycho was aware of the optical writings of Ibn al-Haytham (965–1040) and Witelo (b. ca. 1230), reprinted during his lifetime in Friedrich Risner's important *Opticae Thesaurus* of 1572.[50] Both authors claimed to have investigated the refraction of light from heavenly bodies with an armillary sphere.[51] When Tycho noticed (as Walther had) that as the Sun approaches the horizon in the evening it is apparently displaced progressively further upwards, he correctly attributed this effect to refraction. In order to make up correction tables, he now instituted programs to determine the amounts of refraction suffered by the light of heavenly bodies at various altitudes above the horizon. His efforts were concentrated on the two luminaries, whose predicted diurnal paths were compared with the observed paths after the measured positions had been corrected for parallax. In the Sun's case, Tycho used parallax corrections based on a horizontal parallax of 3′0″ at the Sun's mean distance. His lunar parallaxes were very accurate compared to those of his predecessors but still erred by several minutes at various positions.[52]

Not surprisingly, Tycho arrived at three different empirical tables of refraction corrections, one for the Moon, one for the Sun, and one for the fixed stars which could also be used for the planets. Those for the Sun and Moon differed little: the corrections varied from 34′ and 33′, respectively, at the horizon to imperceptible at an elevation of 45°.[53] In the case of the fixed stars, whose positions needed no parallax corrections, he had compared a few of their refractions with those of the Sun at equal elevations and had found a systematic difference of 4½′. He had therefore simply subtracted that amount from his solar refractions at each altitude, arriving at a table of corrections varying from 30′ at the horizon to zero at an elevation of 20°.[54] Tycho did not, however, claim great accuracy for his refraction tables, especially not for altitudes of less than 20°.[55]

The cause of these refractions presented a problem. If, as Ibn al-Haytham and Witelo argued, refraction was caused by the difference in density between the ether and the sublunary elements, it ought to be perceptible from the horizon to the zenith. Yet Tycho found the amount of refraction insensible above 45°. After debating the issue with Christoph Rothmann,[56] he wrote in his *Progymnasmata* that although a refraction occurred as the light of heavenly bodies left the fluid ether and entered the air, these two media were so close in density that the refraction at their interface was very small and imperceptible to astronomical measuring instruments. The refractions he had measured were caused by the vapors emitted by the Earth which accumulated close to its surface. Above an elevation of 45°, he argued, the effects of these vapors were imperceptible.[57]

But how did Tycho explain the need for three different refraction tables? The only hint of a possible explanation for the empirically derived differences is found in his comment on the use of the stellar refraction table for correcting the measured positions of the planets:[58]

> I suppose, moreover, that these same refractions of the fixed stars can also be applied, not unsuitably, to the planets, except that the Moon is subject to those refractions which we have set forth for the Sun, or perhaps a little greater than these because of its proximity.

Although he had found the lunar refractions to be somewhat smaller than the solar ones, Tycho argued here that they should perhaps be somewhat larger since the Moon is closer than the Sun. He evidently thought that the amount of refraction varied with the distance of the light source![59]

Regardless of the explanation, Tycho's determinations of refractions were a very important addition to positional astronomy. In the case of the Sun, they led to an immediate correction of the obliquity of the ecliptic from Copernicus's value of 23°28′ to 23°31½′.[60] Tycho even sent an assistant to Frombork to verify that Copernicus had indeed not taken refraction into account in taking his solar elevations.[61] Buried in Tycho's empirically derived refraction tables, however, lay the problem of solar parallax.

The relationship between refraction and solar parallax became increasingly important as the seventeenth century progressed.

If Tycho's approach to refraction was long on data and short on theory, Kepler was just the successor to mend that problem. In his *Astronomia pars Optica* of 1604, Kepler put the study of refraction on a firmer theoretical footing. He argued that refraction occurs at the spherical interface between the ether and the air, that refraction continues right up to the zenith, and that it is the same for all heavenly bodies.[62] Using his own law of refraction, which gave excellent results for elevations above 10°, Kepler worked out a single table of atmospheric refraction corrections.[63] When he finally issued the *Rudolphine Tables*, however, he reproduced Tycho's refraction tables, although he did indicate his disagreement and the reasons for it.[64]

* * *

Kepler was, of course, aware of the connection between the problems of refraction and solar parallax, and the subject arose in his correspondence with Remus. Like Kepler, Remus had studied eclipse observations and these had led him to reject Tycho's solar parallax. Remus went further than Kepler, however, concentrating on the relationship between refraction and solar parallax: not only did a very small solar parallax give more consistent results in the study of eclipses, it also went a long way toward erasing the difference between Tycho's solar and stellar refractions.[65]

When Remus visited Kepler in Linz in 1618 and supplied him with the observations of planetary diameters which Kepler used in the *Epitome*, he agreed with Kepler that the solar distance had to be at least 3,000 e.r. and the solar parallax at most 1′.[66] A year later he announced that he had arrived at a solar distance of no less than 6,543 e.r., corresponding to a parallax of about ½′. This figure made the ratio of the Moon's epicycle to its deferent equal to the ratio of the Moon's orbit to the Earth's orbit. It also led to a smaller obliquity of the ecliptic and the need for revising Tycho's refraction tables.[67]

Moreover, while Kepler held fast to the scheme of sizes and distances put forward in his *Epitome*, Remus changed his mind. In 1628 he apprised Kepler of a new scheme, again starting his argument with the problem of refraction. The excess of solar refractions over those of the fixed stars was evidence of an excessive solar parallax. If the angles of incidence of rays of light coming from the Sun and from a fixed star were the same, then the angles of refraction ought to be the same.[68] Remus proposed simply to make the solar refractions equal to those of the fixed stars. Most of the difference of 4½′ could be explained by making the solar parallaxes very small; he would ignore the rest of the difference.[69]

Remus arrived at his solar parallax and distance by an argument similar to Kepler's argument in the *Epitome*. His basis was, however, the new telescopic observations made by himself and Johann Baptist Cysat (1586–1657), Scheiner's student and now himself a professor of mathematics:[70]

> . . . on 4 June 1623 . . . the diameter of Mars observed by Cysat was contained seven or eight times in the aperture of his tube. Now, the tube took in 7′, that is to say, one-fifth of the Moon's diameter, and therefore Mars' diameter was 54½″. (And Cysat adds: "recently it clearly appeared to me to equal the disk of Jupiter or to exceed it a little. Therefore the diameter of Mars is about 1′.") Moreover, Mars was then at 12°58⅔′ of Capricorn, and its distance from the Earth was 93,960, where the semidiameter of the Sun's eccentric is 100,000.

While Kepler had first preferred a proportionality between the planetary surface areas and their distances from the Sun, Remus had convinced him in 1618 that the proportionality should be between their volumes and distances (see p. 85). Now, however, Remus argued for the proportionality between planetary diameters and distances:[71]

> If, however, we assume that Mars' diameter contains 1½ Earth diameters, as is required by the analogy and the proportion of the orbs, and that it subtends 52½″, Mars must be 11,557 semidiameters from the Earth and, consequently, the Sun must be 12,300 semidiameters from the Earth. The parallax of the Sun will therefore be only 17″, and the diameters of all planets will subtend equal angles [seen from the Sun] at mean distances, namely 34″ or (by another adjustment) 36″, which is in wonderful agreement and especially with the harmonies.

The telescopic observations of Mars had given Kepler trouble in the scheme of sizes and distances presented in the *Epitome*. Now Remus had taken the telescopic estimates of Mars's apparent diameter as his starting point and arrived at the proportionality of planetary diameters and distances. He went on to show that observations of other planets did not seriously disagree with this proportionality. If all planets subtend angles of 34″ as seen from the Sun, then the Earth's radius subtends 17″, and the horizontal solar parallax is 17″. This made the Sun's distance 12,300 e.r., an unheard-of number![72]

It should be noted that the telescopic information about the apparent planetary diameters was sufficiently vague and open to question to allow the differing schemes of Kepler and Remus. More important, whereas Remus's scheme was based on mathematical harmony alone, Kepler's was a part of a larger system of mathematical proportions and physical causes, tested at every point against observations. Although in retrospect Remus was closer to what we now consider acceptable values for planetary diameters, his bold leap carried him across that fine line that separates sound scientific practice from mere speculation. Kepler's solar parallax, the foundation of his whole system of sizes and distances, was the result of two decades of studying Tycho Brahe's observations, and it represented, in his considered opinion, the limit of accuracy of the best observations. That limit was the line up to which one could make scientific statements that could be supported by observations; beyond it lay the realm of pure speculation.[73]

In his response to Remus in March 1629, Kepler tried to rein in Remus's enthusiasm. On the one hand, he warned him not to be too dogmatic in putting forward his scheme, for the telescopic observations called both his own and Remus's scheme into question. On the other hand, he asked how Remus's scheme could explain the sesqui-alter proportion between the planetary distances and periods. Moreover, the corrections for the obliquity of the ecliptic and other parameters of solar theory suggested by Remus lay beyond the limits of accuracy of the measurements.[74]

Kepler's achievement was a coherent scheme of sizes and distances based on physical principles, mathematical proportions, and the most accurate observations available. The observations were of two kinds: the positional measurements made by Tycho and the telescopic observations of the planets. By the end of his life Kepler was aware, however, that the latter did not, in fact, support his scheme. If his followers were to ignore Kepler's physical arguments in dealing with the sizes and distances of the planets, his mathematical approach exerted a powerful influence in the seventeenth and even into the eighteenth century. As long as the planetary sizes could not be directly deduced from the phenomena, that is, from accurate measurements of apparent diameters and a direct determination of one absolute distance, astronomers would often resort to the type of harmonic argument used so effectively by Kepler.

* 9 *

Gassendi, Hortensius, and the Transit of Mercury of 1631

At the time of Kepler's death in 1630, there were three schemes of sizes and distances to choose from: Ptolemy's, Copernicus and Tycho Brahe's, and Kepler's. Kepler's scheme, by far the most radical of the three, was based on considerations very different from traditional ones. On the one hand, it was firmly rooted in practical considerations such as telescopic observations and attempts to improve solar theory by means of better corrections for refraction and parallax. On the other hand, its unifying rationale, as it was put forward in book 4 of the *Epitome*, was one of physical causes and mathematical harmonies. In determining the sizes of the planets, Kepler had relied to some extent on telescopic observations of their apparent diameters but to a much larger extent on a pleasing proportion, that is, the proportionality of their heliocentric distances and volumes. As better telescopic observations slowly became available, they progressively limited the scope allowed harmonic speculations. Curiously enough, however, the very first dramatic measurement of an apparent planetary diameter, Gassendi's effort during the transit of Mercury of 1631, actually supported the mathematical speculations of those who followed Kepler in this respect.

Of all the planets known in the seventeenth century, Mercury was by far the most difficult to observe. It was very small yet very bright, and it is never farther than about 23° from the Sun. While in 1610 Galileo had verified the phases of Venus, he was never able to do the same for the phases of Mercury, nor was he able to make a telescopic estimate of its angular diameter. In his *Dialogue* he wrote:[1]

> In Mercury no observations of importance can be made, since it does not allow itself to be seen except at its maximum angles with the sun, in which the inequalities of its distances from the earth are imperceptible. Hence such differences are unobservable, and so are its changes of shape, which must certainly take place as in Venus. But when we do see it, it would necessarily show itself to us in the shape of a semicircle, just as Venus does at its maximum angles, though its disc is so small and its brilliance so lively that the power of the telescope is not sufficient to strip off its hair so that it may appear completely shorn.

Galileo had finished writing the *Dialogue* early in 1630, but it did not, finally, appear in print until two years later. By that time, Mercury had, in fact, been observed shorn of its hair in a dramatic revelation. The possibility of Venus and Mercury passing between the Earth and the Sun had been discussed by Ptolemy in connection with the order of the planets in both the *Almagest* and the *Planetary Hypotheses* (see pp. 20, 21). Transits had, therefore, been discussed in this context in astronomical works, such as Copernicus's *De Revolutionibus* (see p. 47). As mentioned earlier (p. 64), a dark spot seen on the Sun in the year 807 had, in fact, been interpreted as Mercury.

Kepler himself had attempted to witness a transit of Mercury in 1607 by means of a camera obscura, and he had seen what turned out to be a sunspot. Five years later, when the telescope held out hope that transits could be observed, Galileo had criticized Scheiner for trying to observe Venus on the Sun *during a superior conjunction* (see fig. 11, p. 69),[2] but he paid no further attention to the prospects of transits, perhaps leaving his reader with the idea that transits would be invisible even with the telescope. Kepler, however, never abandoned the hope of observing transits (at inferior conjunctions, of course), because he felt that the telescope would surely reveal Venus and Mercury when they crossed the Sun's disk.

In preparing ephemerides for the years 1629 to 1636 from the *Rudolphine Tables*, Kepler and his assistant Jacob Bartsch found that in November 1631 a transit of Mercury would take place, followed a month later by a much rarer transit of Venus, although this latter event would probably not be visible in Europe. In the predictions for the year 1631, Kepler therefore included an "admonition" to astronomers, urging them to observe these two transits. Bartsch made sure that Kepler's admonition would come to the attention of the learned community by printing it separately in 1629, even before the ephemerides themselves had gone to press, and reprinting it in 1630.[3]

Although the calculations showed that the transit of Venus would only be visible in America, Kepler could not rule out small inaccuracies in the time of this event. He therefore directed his request to observe this transit not only to sailors who would be on the high seas, and learned men in America, but also to professors of mathematics in Europe, just in case he had made an error and it would be visible there as well. And he asked them all to observe the Sun on the sixth of December as well as on the seventh, the predicted day for Venus's transit. Mercury's transit was predicted for midday in Europe on the seventh of November, but since the elements of this planet were even more insecure because of the lack of suitable observations, Kepler advised the astronomers to begin their vigil on the sixth of November and, should it go unrewarded, not to stop until the eighth.[4]

Although Kepler mentioned that Venus's parallax would be 4 times as great as the Sun's, and Mercury's 1½ times as great, at the times of their transits, he did not suggest that these events should be used to determine the actual values of the parallaxes of the two planets. Instead, he saw the transits primarily as opportunities for determining the apparent diameters of Venus and Mercury—information that might have important consequences. He reviewed the unsatisfactory state of knowledge about planetary sizes, characterizing the scheme he himself had put forward in the fourth book of the *Epitome* (see pp. 82–90) as based only on probable principles. He even admitted that Mars's apparent diameter as revealed by the telescope did not fit his scheme. However, Remus's notion that the planetary diameters varied as their distances from the Sun would only work if the absolute distances in the solar system were 4 times as great as Kepler himself had made them, and more than 10 times as great as Copernicus had made them. Perhaps the proportions could be saved by varying the densities of the planets in such a way that the predicted apparent diameters would agree with the observed ones.[5]

According to Kepler's calculations, Venus should appear as a round dark spot with a diameter of 7′6″ (almost a quarter of the Sun's apparent diameter) on the Sun's disk during its transit. He had shown this in a sample calculation in the *Rudolphine Tables*, a calculation based on the proportionality of planetary volumes and heliocentric distances. Since Venus's mean distance from the Sun was 0.72 a.u., its volume was 0.72 earth volumes according to this proportionality. Its actual diameter was, therefore, 0.9 earth diameters. Since Kepler made the Sun's horizontal parallax 1′, the Earth's diameter subtended an angle of 2′ from a distance of 1 a.u. At the time of the transit Venus would be 0.26 a.u. removed from the Earth. Seen from the Earth, its apparent diameter should then be $2' \times 0.9 \times 1/0.26$, or about 7′.[6] Kepler did not predict Mercury's apparent diameter during its transit, but the same proportionality and Mercury's distance from the Earth at that time would produce an apparent diameter of about 2½′.

These large predicted apparent diameters led Kepler to a serious misjudgment. He advised astronomers that they would be able to observe the transits not only by projecting the Sun's image through a telescope, but also by projecting it through a simple pinhole. He probably had in mind a rather large camera obscura, such as the one of Philip, Landgraf of Hesse, but he also mentioned his own observation of 1607, when he thought he had observed Mercury on the Sun by simply holding a piece of paper in the beam of sunlight admitted into a dark attic through a small hole in the roof (see p. 64).[7]

Thanks to Bartsch's care, the astronomical community was alerted to the impending transits in good time, and a number of observers made preparations to witness them. Very few, however, were fortunate enough to observe anything. The transit of Venus was, in fact, not visible in Europe, and the transit of Mercury was missed by all but three observers because of cloudy weather or inappropriate instruments. The three astronomers who were fortunate enough to witness this important celestial event were Remus, now in Ruffach in Alsace, his correspondent Cysat in Ingolstadt, and Pierre Gassendi (1592–1655) in Paris. (Kepler himself had died on 15 November 1630). All three projected the Sun's image through a telescope, not a camera obscura.

Gassendi was no stranger to the observation of sunspots. On a number of occasions he had traced the motion of spots across the Sun's disk by means of the telescopic projection method, and now, in anticipation of the two transits, he set up his apparatus in a room in Paris. The image of the Sun on the paper had a diameter of about 8 inches. He drew a circle of that size and divided its diameters (drawn at right angles to each other) into sixty equal parts, so that he would be able to plot the locations of the planet on the Sun's disk in the way the locations of sunspots were usually plotted.[8] Since the Sun's apparent diameter is about 30′, each division represented about 30″. Gassendi began his vigil on the fifth of November.

It rained that entire day and almost all of the following day. The seventh began little better. Shortly before 9 A.M., however, the Sun appeared through a gap in the clouds and produced a clear image on the paper. Gassendi noticed a small spot on the Sun's disk, but it did not occur to him that this might be Mercury. Kepler's figure for Venus's expected size, as well as the traditional notion as to how large Mercury ought to appear, led Gassendi to expect a spot several minutes in diameter, or about a fifteenth part of the Sun's diameter. He wrote:[9]

> . . . I was far from suspecting that Mercury would project such a small shadow. For such was its smallness that its diameter hardly appeared to exceed half of one of the divisions marked. I thought rather that it was a spot which I had not noticed on the Sun on a previous day.

Whereas earlier observers had believed that they were observing Mercury when, in fact, they were looking at sunspots, Gassendi now thought that he was seeing sunspots when, in fact, he was observing Mercury on the Sun!

As the clouds parted intermittently, Gassendi was able to observe the Sun from time to time. He measured the position of the small spot in the hope of using it as a reference point. Such a spot, sharing the Sun's 26-day axial rotation, would show no sensible change of position over a period of hours. Mercury, on the other hand, would move across the entire disk of the Sun in a matter of a few hours, should it appear. After two measurements, however, Gassendi found to his surprise that this dot, this supposed sunspot too minute to be Mercury, was moving much too rapidly to be a sunspot:[10]

> Thereupon, thrown into confusion, I began to think that an ordinary spot would hardly pass over that full distance [of four divisons] in an entire day. And I was undecided indeed. I could hardly be persuaded, however, that it was Mercury, so much was I preoccupied by the expectation of a greater size. Hence I wondered if perhaps I could have been wrong in some way

> about the distance measured earlier. And when the Sun shone again, and I ascertained the apparent distance to be greater [again] by two divisions . . . , then at last I thought that there was good evidence that it was Mercury.

Obviously, Kepler's statement about the expected size of Venus during its transit was not the only factor that had led Gassendi astray in his expectation of Mercury's size. He had, no doubt, been taught the traditional, canonical scheme of sizes and distances during his university education, and although as a Copernican he came up with distances somewhat different from those of Ptolemy and his followers, when it came to the apparent sizes of heavenly bodies he agreed with them entirely. Gassendi, then, was expressing here a deeply ingrained expectation of planetary sizes, and this was two decades after the telescope had entered astronomy!

Convinced, now, that he was actually observing Mercury on the Sun, Gassendi compared the diameter of this little dark spot with the diameter of the Sun and found that it "could scarcely exceed two-thirds of one division, that is, the third part of a minute, or 20 seconds."[11] This was only about a sixth of the apparent diameter which Tycho's figures and Kepler's admonition would lead one to expect.

In his report, *Mercurius in Sole Visus et Venus Invisa*, published as an open letter to his colleague Wilhelm Schickard (1592–1635), Maestlin's successor in the chair of astronomy at Tübingen, Gassendi shared his astonishment with the astronomical community. Although others, especially Galileo, had pointed out that, when observed with the naked eye, the stars and planets are clothed in adventitious rays and appear smaller when stripped of these rays by the telescope, no one had yet seen this very bright and very small planet well defined. Now, Gassendi had seen Mercury stripped naked and starkly outlined against the Sun's disk. Referring rhetorically to the mythical figure of Hermes Trismegistus, he asked "who had, in fact, persuaded himself that Mercury could be thrice great on Earth and that he could now appear thrice small in the heavens?"[12]

Cysat only reported his measurement of Mercury's apparent diameter to his correspondents. He had determined it to be 25″, but he argued that since the Sun is much larger than Mercury it illuminates more than half the planet's globe, so that the observed dark spot was smaller than the apparent disk of Mercury.[13] He did not say how much smaller, but he could have profited from reading Aristarchus's *On the Sizes and Distances of the Sun and Moon*, written two millennia earlier, in which Aristarchus had shown (in the case of the Moon) that the difference between the illuminated part and a complete hemisphere is negligible (see p. 8). Although he did not say so, surely we may assume that Cysat grasped at this straw because he had difficulty accepting such a small apparent diameter for Mercury.

Remus, on the other hand, measured Mercury's shadow to be only 18″, and he accepted this value as the apparent diameter of Mercury without hesitation.[14] Small wonder, for it fit his scheme of planetary sizes very neatly. If seen from the Earth Mercury had an apparent diameter of 18″ at inferior conjunction, then seen from the Sun it would have an apparent diameter of about 30″. But if one assumed, with Remus, a solar distance of 12,300 e.r., the Earth and Mars also had apparent diameters of 30″ seen from the Sun. The apparent diameters of all planets seen from the Sun were therefore the same, and their actual diameters were in proportion to their heliocentric distances.[15] Remus had proposed this proportion in 1628 on the basis of his measurement of Mars's apparent diameter (see p. 93), and now it had been confirmed by his measurement of Mercury. Would Kepler have abandoned his own scheme in favor of Remus's had he still been alive?

For the time being Gassendi's observation of Mercury was the only one to find its way into print. His surprise at the planet's smallness was shared by his colleagues. Only Galileo received the news without surprise.[16] Claude Nicholas Fabri de Peiresc (1580–1637), a patron and the center of a scientific circle in the south of France, informed Gassendi that the astronomer Joseph Gaultier de la Valette in Aix-en-Provence had observed diligently all day on the seventh of November under blue skies but had failed to see anything on the Sun.[17] The Landgraf of Hesse and his staff had also observed all day without seeing anything. Led astray by Kepler, they had used the Landgraf's camera obscura.[18]

Wilhelm Schickard, to whom Gassendi's tract was addressed, had also been led astray. Following Kepler's example of the pretelescopic observation of what turned out to be a sunspot in 1607 (see p. 64), he had made a hole in one of the tiles of his roof, and with some of his friends he had awaited the glorious event in his attic. They had, however, been frustrated by the weather: it was cloudy all day, and the Sun hardly poked through the clouds once or twice.[19] Had the Sun shone all day, they would have been equally frustrated!

In his printed reply to Gassendi, Schickard expressed his surprise at Mercury's "entirely paradoxical smallness" and gave his opinion that Gassendi would "hardly escape the censures of the critics, who will doubt from this whether you have really seen Mercury himself."[20] Since, however, the motion of the spot proved that it could not have been a sunspot, Schickard agreed that it had to be Mercury passing across the Sun's disk. Tycho and Kepler had, however, predicted a much larger apparent diameter, and Schickard therefore set himself the task of explaining away Mercury's observed paradoxical smallness.

After the telescope had stripped away the extraneous rays that appear to surround Mercury, Schickard argued, the remaining disk should have a diameter of about 1′. It appeared smaller, however, for three reasons. A stick held closely in front of a flame appears thinner from a distance than it is in reality, because it is the nature of light to spread itself in all directions. Likewise, Mercury's disk, surrounded by the brillance of the Sun, would appear smaller than it really is. Furthermore, as Cysat also argued, the Sun, being so much larger, illuminates more than half of Mercury's globe, leaving a smaller unilluminated side visible from the Earth. And, last, perhaps Mercury was not entirely opaque but had a solid core surrounded by a translucent coat.[21] This last farfetched argument to preserve Mercury's expected, canonical size had already been used in 1610 by Lodovico delle Colombe (1565–ca. 1615), leader of the opposition against Galileo in Florence, in a vain effort to preserve the perfection of the Moon in the face of telescopic evidence about its mountainous nature.[22]

The Landgraf's observation, however, would tend to support Gassendi's observation. Schickard pointed out that the Landgraf's celebrated camera obscura could not reveal sunspots with diameters of less than 2′. He concluded that although Gassendi's observation itself was correct, Mercury's smallness was only apparent. In reality its apparent diameter was considerably larger, that is, about 1′. For various optical reasons, however, Mercury showed itself as a spot of only about 20″ on the disk of the Sun.[23]

If Cysat and Schickard tried to explain Mercury's smallness away, Martinus Hortensius (1605–39), their young colleague in Amsterdam and a disciple of Philip van Lansberge (1561–1632), accepted Gassendi's observation at face value. Hortensius had made thorough preparations to observe both transits by means of the telescopic projection method. On both occasions, however, he had been foiled by the weather. Since he had been interested in the problem of planetary sizes for some time, he took this opportunity to publish a tract on the subject. His *De Mercurio in Sole Viso* appeared in 1633.

Although he was a member of the first generation of astronomers who had grown up with the telescope, Hortensius was by no means unfamiliar with the traditional expectation of planetary sizes. He therefore understood Gassendi's difficulty in convincing himself that he was really observing Mercury on the Sun:[24]

> It is indeed like that: how often we are held by prejudice, so that we either do not admit what is before our eyes or bow as much as possible to a preconceived opinion because of the perverse wont of human nature. Nor do I except myself from that weakness: I would have thought the same, had I succeeded in observing Mercury.

He did not, however, try to preserve the traditional figures for the apparent diameters of the planets. Instead of trying to explain how Mercury could in reality be larger than Gassendi had measured it, he demolished Schickard's ill-chosen optical arguments and accepted Gassendi's measurement.[25]

After disposing of Schickard's arguments, Hortensius turned his attention to the apparent diameters of the other planets and the fixed stars. He himself had tried to make measurements, or at least estimates, of apparent diameters, but, as is abundantly clear from his text, this was a difficult business. By observing two fixed stars of known separation, close enough to each other so that they could both be brought into the field of the telescope at the same time, he had estimated the diameter of the field of view of his telescope. He had then been able to estimate what portion of the field was occupied by Jupiter, and he concluded that Jupiter's apparent diameter was no more than 1′. The observation of Mercury's transit, however, gave him the idea of trying to project Jupiter's image on paper, and on 13 November 1632, when Jupiter was at opposition, he had actually managed to do this. Comparing the diameter of Jupiter's image with the diameter of the Sun's image projected through the same telescope with the paper held at the same distance behind it on the previous and following days, Hortensius obtained ratios of 6:195 and 4:130. Using Lansberge's figure of 34′ for the Sun's apparent diameter on that day, Hortensius concluded that at opposition Jupiter's apparent diameter was about 60″.[26]

These measurements were exceptional, however. After failing in his attempts to project the images of Mercury, Mars, and Saturn, Hortensius concluded that Venus was the only other planet bright enough to allow this procedure. So far, he had missed several good opportunities. However, he had compared the apparent diameters of Venus and Mercury during a conjunction, and that same evening he had compared the diameter of Venus with that of Jupiter. He found Venus to be ⅓ as large as Jupiter and Mercury ⅔ as large as Venus. Knowing from the previous measurement of Jupiter that on that evening its angular diameter was about 48″, he concluded that the apparent diameters of Venus and Mercury were about 16″ and 10″, respectively, that night. When he had adjusted for Mercury's distance, he found that his roundabout estimate agreed closely with Gassendi's measurement made during Mercury's transit. Hortensius had also measured Mars's apparent diameter by comparing it with the Moon's during a conjunction of these two bodies in 1632, and he had compared Saturn's diameter with Jupiter's.[27] Assuming, with his mentor Lansberge, a solar distance of about 1,500 e.r., Hortensius now drew up a table of apparent and actual planetary sizes (see table 13).[28]

In comparing Hortensius's figures with pre-telescopic values for apparent diameters, one is struck not only by the differences in magnitude, but also by the fact that Hortensius gave apogee and perigee values, whereas before the telescope only the apparent diameters at mean distances had been given. Galileo had already drawn attention to the great variations in the apparent sizes of Venus and Mars and had argued that this was evidence for the Copernican system. Regardless of the merit of this argument, with the telescope it became customary to measure the apparent diameter of a planet and then, knowing its position and relative distance from the Earth, to calculate its apparent diameters at apogee and perigee, and often also at mean distance. Hortensius's scheme was the first to be cast in this form.

TABLE 13 Hortensius's Planetary Sizes

Planet	Apparent Diameter at Apogee	Apparent Diameter at Perigee	[Actual Diameter] [Earth = 1][a]	Volume (Earth = 1)
Mercury	10″	28″	1/18.7	1/6510
Venus	15⅓″	1′40″	1/10.35	1/1109
Mars	9″	1′4″	1/11.53	1/1534
Jupiter	38½″	1′1⅔″	1/1.08	1/1.25
Saturn	31″	42⅓″	1.30	2⅕

[a]Not given by Hortensius.

When it came to the diameters of the fixed stars, the telescope was less helpful because it did not reveal disks. Hortensius estimated Sirius's angular diameter to be 10″, while the rest of the fixed stars had apparent diameters ranging from 8″ for the first magnitude to 2″ for the sixth. Whereas Lansberge had assumed an annual stellar parallax of 15″, which would make the distance to the fixed stars about 28,000 a.u., and Sirius's actual diameter 1⅓ a.u., Hortensius accepted Tycho's upper limit and assumed an annual stellar parallax of 1′, which corresponds to a stellar distance of 6,875 a.u. Sirius's actual diameter was thus about ⅓ a.u., while a fixed star of the sixth magnitude had a diameter of about 1/15 a.u., or about 100 e.r.[29] In this manner Hortensius tried to make the fixed stars somewhat smaller than his mentor had made them. Within the sphere of Saturn, Hortensius's assumptions made the Earth about equal to Jupiter in size and somewhat smaller than Saturn, but Mars, Venus, and Mercury were much smaller than the Earth.

Kepler's admonition had served to alert observers to a phenomenon never before observed in the heavens. In their attempts to witness these transits astronomers were aware, of course, that their observations could provide excellent opportunities to check the times and positions predicted in the tables and to improve the parameters of the planetary models of Mercury and Venus. This was especially important in the case of Mercury, the most difficult planet to observe. They were also not unaware that these transits could perhaps serve to determine the parallaxes of Mercury and Venus, provided that suitable observations could be made at different locations and compared. For both reasons Gassendi made preparations to have an assistant measure the Sun's altitude with a quadrant (in order to determine the time) in a room directly below him each time he gave the signal by stamping his foot. But the young man had gone out, convinced that the clouds would not allow the observation to be made, and by the time he had been found there was time for only one measurement before Mercury had left the Sun's disk.[30]

Accustomed as we are to thinking about transits primarily in connection with parallax measurements, we must be careful not to make the mistake of thinking that this measurement or even the correction of planetary elements, was considered by Gassendi and his colleagues to be the most important aspect of Mercury's transit of 1631. From Kepler's admonition to Hortensius's defense of Gassendi, the main issue was Mercury's apparent size. For Kepler this measurement was of crucial importance for his scheme of sizes and distances. Although corrections were made in Mercury's elements as a result of Gassendi's observation, no parallax of Mercury resulted from it. The planet's "entirely paradoxical smallness" was by far the most important result of the observations of the transit of 1631.

In principle this smallness was implicit in Galileo's various statements about planetary sizes, from *Sidereus Nuncius* of 1610 to his *Dialogue* of 1632. In the absence of a table of measurements to take the place of the traditional measures, however, the full meaning of Galileo's general utterings on this subject had been lost on scientists in general and even astronomers in particular. With the exception of Galileo and perhaps Remus, astronomers, even those like Gassendi who regularly scrutinized the heavens through a telescope, had not rid themselves yet of the traditional notion of how large planets and fixed stars ought to appear. Gassendi's announcement that Mercury's apparent diameter was an order of magnitude smaller than almost anyone had supposed it to be provided an essential and dramatic illustration of Galileo's general point, so vigorously reiterated that same year in the *Dialogo*. Gassendi's tract was sufficiently important to elicit two published replies, both dealing primarily with Mercury's unexpected smallness.

The need for a new set of apparent sizes, based entirely on telescopic measurements (or at least estimates), was now incontestably evident. Hortensius had made these difficult measurements and estimates after reading Gassendi's tract and becoming aware of the importance of this project. His table of angular diameters (table 13) was the first such scheme based entirely on information gathered by means of the telescope. It remained the only one for nearly two decades. With the possible exception of Tycho Brahe, Hortensius was the first astronomer since Hipparchus to determine the apparent sizes of the fixed stars and all the planets entirely for himself without being influenced by received opinion.

* 10 *
From Horrocks to Riccioli

In the generation following Kepler's death, telescopic astronomy gradually became a science in its own right, one with a continuous, not spasmodic, research tradition. Yet, in the absence of the micrometer and telescopic sight, telescopic studies tended to remain qualitative. Knowledge of sizes and distances, therefore, remained approximate and speculative for some time, with the exception of the apparent diameters of Mercury and Venus, measured with accuracy during transits in 1631 and 1639. As the eclipse diagram was more and more discredited, two methods of trying to find absolute distances remained: Aristarchus's lunar dichotomy and Kepler's harmonic speculation. Of these, the followers of Kepler chose the latter, although they did not agree with the particular proportion he had favored. Those who did not share Kepler's Copernican convictions or his speculative tendencies had little choice but to try to perfect Aristarchus's venerable method. In both cases the connection between solar theory and corrections for solar parallax and refraction was becoming more central to the problem of sizes and distances.

The publication of the *Tabulae Rudolphinae* or *Rudolphine Tables* in 1627 was the crowning glory of Kepler's long efforts. The tables were based on Tycho Brahe's observations, much more accurate than any made before, and Kepler's elliptical planetary theories, which better represented the phenomena than the eccentrics and epicycles of his predecessors. Because of their unprecedented accuracy, these tables became the new standard, making all previous tables obsolete. Thus, they allowed Kepler to predict the transits of Venus of 1631 and 1761. In this latter prediction he erred by only two days.[1]

That the *Rudolphine Tables* were by no means perfect is illustrated by the fact that while pinpointing a transit of Venus that would occur 132 years hence, Kepler entirely missed one that was to occur within a decade.[2] When one compared predictions based on the tables with observed positions, small but significant errors in the planetary models remained. One important source of errors was the faulty correction of solar positions for the effects of parallax and refraction. Wrongly corrected positions had led to small errors in the parameters of the model of the Sun's, or rather the Earth's motion, and these, in turn, caused errors in the theories of all the planets. The elimination of this source of errors was the achievement of the brief career of Jeremiah Horrocks (1618–41), and it allowed him to predict the transit of Venus missed by Kepler. Horrocks's observation of this transit in 1639 supported his conviction that the solar parallax was smaller than Kepler had made it, and that the solar distance was almost inconceivably large.

Horrocks and his older friend William Crabtree (1610–44), working in Lancashire, England, were among the first astronomers to accept Kepler's elliptical astronomy and to try to improve it. In 1635, at the age of 17, Horrocks began to compute ephemerides from the *Tabulae Motuum Coelestium Perpetuae* by Philip van Lansberge. Although a Copernican, Lansberge rejected Keplerian ellipses, preferring the traditional eccentrics and epicycles.[3] When Horrocks was converted to Kepler's astronomy by Crabtree in 1637, he compared the *Rudolphine Tables* with the tables of Lansberge and found the former much more accurate.[4] He also found small errors in the *Rudolphine Tables,* however, and he set it as his task to improve the planetary parameters used by Kepler in these tables. For various reasons Horrocks concentrated his efforts on the orbit of Venus.[5]

The values used to correct the Sun's altitudes for parallax and refraction have a direct impact on solar theory. To find the eccentricity of the Sun, the astronomer had to determine the times of the solstices and the equinoxes, and these times were found by measuring the noonday altitudes of the Sun. Missing or incorrect parallax and refraction corrections will result in errors in the times of the solstices and equinoxes, thus yielding an incorrect solar eccentricity. As we have seen, Ptolemy and his successors, up to Copernicus, had corrected solar altitudes for parallax only (ignoring the effects of atmospheric refraction), using a horizontal solar parallax of 3′. The resulting errors led to solar eccentricities (Ptolemy made it 1 part in 24 or 0.417, and Copernicus 1 part in 31 or 0.0323)[6] that were about twice too large, and this error in the Sun's theory also led to errors in the theories of the planets.

Tycho began the practice of correcting his solar observations for refraction. However, his refraction corrections left a great deal to be desired, partly because he retained the horizontal solar parallax of 3′ of his predecessors. As a result, his value for solar eccentricity, 0.03584,[7] was still much too large. In his *Rudolphine Tables,* Kepler reduced the horizontal solar parallax to 1′, but he did not make the concomitant correction in the solar eccentricity. Horrocks uncovered this shortcoming and calculated a new solar eccentricity, 0.0173, from Tycho's observations, using Kepler's value of 1′ for the Sun's horizontal parallax.[8] Agreeing with Kepler that 1′ was the upper limit of the Sun's parallax, Horrocks wished to know its actual value. After a thorough study of the eclipse diagram, he rejected this method;[9] he also found the method of lunar dichotomy wanting.[10] All that was left was Keplerian harmony, but Kepler's proportionality of planetary distances and volumes had to be rejected after Gassendi's determination of Mercury's apparent diameter during the transit of 1631. Horrocks looked for a different proportion.

Horrocks's correction of the solar eccentricity resulted in an important improvement in the parameters of the theory of Venus. In 1639 he found that, according to the improved theory, Venus would pass across the Sun's disk during its inferior conjunction that was to occur in December of that year. In an effort to alert others to this impending event, he wrote to Crabtree on 5 November:[11]

> The reason why I am writing you now is to inform you of the extraordinary conjunction of the Sun and Venus which will occur on November 24 [o.s.]. At which time Venus will pass across the Sun. Which, indeed, has never happened for many years in the past nor will happen again in this century. I beseech you, therefore, with all my strength, to attend to it diligently with a telescope and to make whatever observation you can, especially about the diameter of Venus, which, indeed, is 7′ according to Kepler, 11′ according to Lansberge, and scarcely more than 1′ according to my proportion.

Obviously this transit would present a wonderful opportunity to "fine tune" some of the parameters of Venus's theory, but the young astronomer was also vitally interested in the apparent diameter of Venus during this event, singling it out in his letter. Indeed, he had his own expectations about how large the shadow of the planet would appear, an expectation based on a proportionality. Horrocks had been in possession of a telescope adequate for astronomical observations only since early in 1638 and had made only a few rough comparisons of the apparent diameters of some planets.[12] Moreover, he had not read Galileo's letters on sunspots or Hortensius's *De Mercurio in Sole Viso.*[13] He only knew of the approximate planetary apparent diameters reported in Kepler's *Epitome* and Gassendi's measurement of Mercury, and therefore his proportionality of planetary sizes was based on these only. The exact nature of this scheme will become apparent below.

Convinced that Venus would be much smaller than Kepler had predicted, and having read Gassendi's account of the transit of Mercury, Horrocks knew that the observation could not be made with a pinhole camera. He and Crabtree both used the telescopic projection method for their observations, and although it is usually cloudy in Lancashire at that time of the year, both were lucky enough to witness part of the transit. Venus did not enter the Sun's disk until late in the afternoon on the appointed day, 4 December, and owing to the shortness of winter days in those northern latitudes, the transit was visible for less than an hour before the Sun set. Horrocks observed and measured Venus during this entire period; Crabtree had time only for one rough measurement of the planet's apparent diameter.[14] They were the only ones to observe this transit.

During the first few months of 1640 Horrocks measured the apparent diameters of the other planets as best as he could. He not only compared apparent diameters using Venus as a standard, in the manner of Hortensius, but also used the questionable naked-eye method suggested by Galileo in his *Dialogue* (see p. 75), the Latin edition of which (Strasbourg, 1635) Horrocks had finally obtained in 1639.[15] He looked at a planet through a pinhole and moved an iron needle (instead of a cord) back and forth until he found a distance at which it just covered the planet.[16]

Horrocks incorporated these new measurements into a tract describing his observation of Venus's transit. By December 1640 he was nearing the completion of *Venus in Sole Visa,* and he wrote to Crabtree for advice on how to get it published. He was to have visited Crabtree early in January, but tragedy struck. On the day before the appointment Horrocks died; he was not yet 23. Crabtree himself died at an early age in 1644,[17] perhaps a victim of the Civil War. As a result, *Venus in Sole Visa* and other equally important unpublished writings of Horrocks were known only in a restricted provincial circle in England, from which knowledge of them spread only gradually. They were not published until after the Restoration.[18]

Horrocks had two main concerns when he wrote *Venus in Sole Visa:* the correction of the elements of Venus's orbit, and the determination of the sizes and distances of the planets. The central measurement was, of course, that of Venus's apparent diameter obtained during the transit. During his brief observation, Crabtree had managed to obtain a ratio of 7:200 between the apparent diameters of Venus and the Sun.[19] This supported Horrocks's own measurement, which gave the ratio $1\frac{12}{60}$:30. Given an apparent solar diameter of 31′30″ on that day, this last ratio gave Venus's apparent diameter as 1′16″, much smaller than the 12′18″ predicted by Tycho:[20]

> Congratulate us, Gassendi, on clearing from suspicion your observation of Mercury, and let astronomers cease to wonder at the surprising smallness of the least of the planets, now they find that the one which seemed the largest and brightest scarcely exceeds it. Mercury may well bear his loss since Venus sustains a greater.

Aware of Schickard's response to Gassendi, Horrocks devoted considerable space to countering the possible objections to the smallness of Venus's apparent diameter. In doing so, he made liberal use of Galileo's general arguments on this subject found in the *Dialogue*. Because of his isolation, however, he had not seen the letters on sunspots, which contained Galileo's specific estimates of Venus's apparent diameter (see p. 70)—estimates that confirmed Horrocks's figure.[21]

Having established the validity of his measurement, Horrocks addressed himself to the problem of "the dimensions of the wandering stars, and . . . the horizontal parallax of the Sun, a matter of the greatest importance, and one which has been the subject of much fruitless speculation."[22] This did not mean, however, that Horrocks's account was going to be entirely factual: he made it clear that he, too, had no choice but to speculate. Presumably, then, his speculations were more fruitful than those of his predecessors. While he agreed with Kepler that "the proportion of the globes and orbits [that is, distances] of the planets is most accurate," he disagreed with his choice of measures for the planetary globes: comparing planetary distances with *volumes* is like comparing "the head of one person with the foot of another."[23] Moreover, Kepler's proportionality did not square with the new telescopic evidence:[24]

> It is clear from the example of Venus that experience is entirely against the proportion of Kepler; and this is also evident from Gassendi's observation of the planet Mercury, the diameter of which he found to be scarcely equal to the third part of a minute, although Kepler's calculation extends it to three minutes. The same is the case with reference to Mars whose diameter, according to Kepler's rules, is sometimes increased beyond six minutes; whereas, in reality, it never equalled two: and Kepler himself confesses that when Mars was nearest the Earth, he did not appear much larger than Jupiter which he estimates at only fifty seconds. He errs less, scarcely at all, with regard to Saturn and Jupiter.

Unaware that Remus had already traveled this road, Horrocks now made the *diameters* of the planets proportional to their distances from the Sun. This was the proportion to which he had referred in his letter to Crabtree in 1639 announcing the upcoming transit. Horrocks thus arrived at the proportionality before the transit observation. He could not have found it on the basis of observations alone at any rate, for, as discussed in chapter 8, such a proportionality implies a particular value for the solar parallax. Thus, Kepler had used a solar parallax of 1′, and he had argued on this basis that the proportion put forward by Remus, and now repeated by Horrocks, was impossible. Horrocks, however, disagreed with Kepler's argument:[25]

> But Kepler writes that the proportion of the diameters is without doubt disproved by observation. I reply that he created a shadow which prevented him from seeing clearly. It is true that observation is opposed to it, if his parallax of the Sun, which is of one minute, is to be taken; but I see no necessity for adopting such a parallax, nor do I acknowledge the propriety of his original speculations, much less of his other arguments. Such reasoning is absurd, and like begging the question; the true proportion of the orbits and globes should be sought from observation. In this way the apparent semi-diameter of the Earth, or parallax of the Sun, may be concluded; and if this is borne out by observation the thing is finished.

Now, Kepler's solar parallax was not based on physical and mathematical arguments only. As we have seen, these speculations dovetailed with, indeed they were the justification for, a solar parallax to which Kepler had been driven by his study of eclipses and his examinations of Tycho's attempted measurements of Mars's parallaxes. After Gassendi's measurement of the apparent diameter of Mercury, however, Kepler's proportion was no longer tenable. Horrocks therefore opted for the proportion suggested by Gassendi's measurement of Mercury, supported by his own measurement of Venus, and also not inconsistent with his estimates of the diameters of Mars, Jupiter, and Saturn.[26] Although not as completely interwoven in other parts of his astronomy as Kepler's proportion had been, Horrocks's proportion was nevertheless part of a larger whole. It was supported by the adjustments he had made in the parameters of the theories of the Sun and Venus, and it in turn produced a solar parallax that was consistent with his improvements of these theories. If his youthful enthusiasm led him to overstate his case when he argued that his solar parallax followed directly from observation, the apparent diameters of the planets, seen from the Sun, did seem to be equal:[27]

> Since therefore it is certain that the diameters of the five primary planets, in mean distance, appear from the Sun 0′28″, and that none of them deviate from this rule, tell me, ye followers of Copernicus, for I esteem not the opinions of others, tell me what prevents our fixing the diameter of the Earth at the same measurement, the parallax of the Sun being nearly 0′14″ at a distance, in round numbers, of 15000 of the Earth's semidiameters? Certainly, if the Earth agree with the rest as to motion, if the proportion of its orbit to that of the rest be so exact, it is ridiculous to suppose that it should differ so remarkably in the proportion of its diameter.

Without any assumption about absolute distances, Horrocks could calculate from his own measurements of the planetary apparent diameters what their apparent diameters would be when seen from the Sun. To his own satisfaction, he deduced directly from the phenomena that, seen from the Sun, the apparent diameters of all planets except the Earth were the same: 28″. He could not know, however, what the Earth's apparent diameter as seen from the Sun would be. For this he needed to know the Sun's parallax or absolute distance. For a Copernican the Earth was merely another planet, and so it was entirely reasonable to assume that its apparent diameter seen from the Sun should be the same as those of the other planets. A follower of Tycho or Ptolemy would surely question the grounds for this last step.

Horrocks argued that, although his proportion, like Kepler's, rested ultimately on speculation, his speculation was probable and, at any rate, it was a proper way of arguing. To anyone who might object that his solar distance was unbelievably large, he replied that his observation of Venus had shown how faulty the accepted measures of planetary sizes and distances could be. He admitted finally, however, that in the ultimate analysis observations would be decisive. If future observations were to prove his scheme wrong, he would reject his opinion as false speculation.[28]

Horrocks thus agreed with Kepler that the Earth must be related in size to the other planets by some proportion, but, following unknowingly in the footsteps of Remus, he improved on Kepler's scheme by substituting a proportion that, as he argued, fit the observations better. Except for Gassendi's measurement of Mercury and Horrocks's own measurement of Venus, however, the existing estimates of apparent planetary diameters were very inaccurate. Kepler had doubted that, for purposes of deciding on a particular proportionality, telescopic observations were "certain enough not to be taken exception to."[29] Except for the transit observations of 1631 and 1639, nothing had changed since Kepler wrote this in his *Epitome*.

Although Horrocks had been informed in 1639 that William Gascoigne (1612?–44) had devised an instrument for measuring small angular distances with a telescope,[30] he had no access to such an instrument, and therefore he could only estimate the angular diameters of the superior planets in much the same way as Galileo, Kepler, Remus, Cysat, and Hortensius had done. He had estimated Mars's angular diameter at opposition to be 2′ or less, Jupiter's diameter "by twilight" (that is, presumably near conjunction with the Sun) to be 37″, and Saturn's to exceed half a minute but not to equal a whole minute.[31] Obviously, such estimates could be used to confirm a wide range of proportions. It is also clear from his letter to Crabtree that Horrocks had already determined his proportion before the transit of Venus, indeed before he had made any significant estimates of apparent planetary diameters at all.[32] He, therefore, must have arrived at it by examining Kepler's argument in book 4 of the *Epitome,* knowing also the result of Gassendi's observation of Mercury on the Sun. The proportion was thereupon neatly confirmed by his own observation of Venus on the Sun.

Strictly speaking then, Gassendi's and Horrocks's own transit observations showed only that at their mean distances Mercury and Venus subtend roughly equal angles as seen from the Sun. Horrocks's measurements of the superior planets hardly confirmed this proportion in an unambiguous way. The proportion and the solar parallax derived from it, therefore, had to be treated with a certain amount of skepticism, and Horrocks admitted as much. In *Astronomia Kepleriana Defensa & Promota,* a work later pieced together by John Wallis (1616–1703) from various pieces written by Horrocks between 1638 and his death, the discussion of this subject shows Horrocks in a more careful mood. He gave the solar parallax as 15″ in one place and 14″ in another,[33] and he expressed his opinion of the accuracy of these figures as follows:[34]

> You will understand, kind and understanding reader, that I do not define the distance of the Sun in feet and inches, which nearly all other astronomers do, since they are afraid to omit

> one-sixtieth part of a semidiameter lest, forsooth, something of great importance be lost. For my part, I could not attain such subtlety (or rather vanity) but rather I demonstrate in round numbers a very great distance, outstripping all senses. I shall, therefore, not consider someone to dissent from me sensibly if he adds to or subtracts from the distance of the Sun even a thousand semidiameters of the Earth. And the same is to be said much more of the size, with which, if you were to double it, you would not find me disagreeing very much. For a small and imperceptible change in the Sun's parallax would produce all this and more. All I undertake to demonstrate in this treatise, moreover, is this, that the horizontal parallax of the Sun is almost insensible and much smaller than is generally estimated by astronomers, and, finally, that the astronomical observations put nothing in the way of, nay rather they confirm that parallax which I determined by very probable conjectures in the book about Venus seen in the Sun. This I affirm. Of him who gathers more from my opinion I ask that he offer it entirely as his own and not bring me undeserved odium.

Unfortunately neither this work nor *Venus in Sole Visa* saw the light of day until much later. For the time being, little new was added to the subject of planetary sizes and distances. Gassendi made no measurements of angular diameters to speak of, and in his influential textbook, *Institutio Astronomia,* of 1647 he reported the traditional Ptolemaic scheme as well as those of Tycho and Hortensius. But he did express his preference for Hortensius's results.[35] Although Pierre Hérigone (Clément Cyriaque de Mangin, d. 1642) reported Hortensius's scheme in his all-encompassing *Cursus Mathematicus* (1634–37), he preferred Tycho's scheme.[36] In his mystical *Oculus Enoch et Eliae* of 1645, the Capuchin monk Antonius Maria Schyrlaeus de Rheita (1597–1660), a defender of Earth-centered astronomy, assumed that the Sun's diameter was 10 times as great as the Earth's—a solar parallax of 1′30″—and that the planetary diameters were proportional to the square roots of the periods.[37] And Johannes Hevelius (1611–87), a Copernican and the undisputed master of telescopic astronomy, reported only one or two primitive measurements of apparent diameters in his sumptuous *Selenographia* of 1647.[38]

The one rather daring departure from orthodoxy during the 1640s came from the Belgian astronomer Gottfried Wendelin (1580–1667), an outspoken Copernican despite his priestly vows. In *Loxias seu de Obliquitate Solis Diatriba* of 1626, Wendelin had actually followed Kepler's advice to try to determine the solar distance by Aristarchus's method of lunar dichotomies. He found that at the time of dichotomy the Moon was at most 1° removed from quadrature, so that the solar parallax was at most 1′.[39] In a letter to Gassendi in 1635 Wendelin argued that the planetary diameters were in the ratio of their distances from the Sun, so that from the Sun all planets, including the Earth, subtended an apparent diameter of about 28″ or at most 30″. This meant a solar parallax of at most 15″ and a solar distance of 14,720 e.r.[40]

When in 1644 he published his analysis of lunar eclipses from 1573 to 1643, Wendelin took a solar distance of 14,656 e.r. This was the first suggestion in print that the solar distance was greater than Kepler had made it.[41] When the learned Giovanni Battista Riccioli (1598–1671), engaged in his review of the entire body of astronomical knowledge, read *Eclipses Lunares,* he was surprised at this enormous solar distance and wrote Wendelin a letter asking his reasons for it. Riccioli reported Wendelin's reply as follows:[42]

> He replied, at last, on 9 November 1647, that he had done it because a solar parallax of at most 15″ better represented the equinoxes, their intervals, and eclipses; and because it better reconciled the altitude of the Pole, taken from the circumpolar stars at Forcalquier in Provence, with the distance between the tropics found at the solstices; but most of all because of lunar dichotomies. For although formerly it had appeared to him that when the Moon appears to us halved it was distant from the Sun 89°, and that therefore the Sun was removed from us about sixty times the lunar distance, that is, about 3,600 terrestrial semidiameters, he says, "afterwards, however, taking greater care and with better telescopes, I discovered . . . that the dichotomy occurs at a distance of about 89½° between the Moon and Sun, from which are found 7,000 semidiameters already, more or less. But finally, having compared the dichotomies with each other, the morning as well as the evening ones, to consider how much was conceded by the unevenness of the Moon, and having considered also that the light of the Sun entering into the lunar air advances the evening dichotomy and retards the morning one, with all the circumstances considered, I could not decide other than that the dichotomy occurs [*refringere*] at a distance of 89°45′, or even more. From which followed a distance between the Sun and the Earth of at least 13,740 semidiameters, which gave excellent agreement with the above examinations by means of parallaxes." After this, having considered the intervals of the planets, and also of the jovial ones, and their admirable harmony, about which elsewhere, he concluded that the mean distance of the Sun from the Earth was 14,656 terrestrial semidiameters.

Obviously Wendelin had found from his analysis of eclipses that a very small solar parallax gave better results than the larger values adopted by Ptolemy, Copernicus, Tycho, and even Kepler. And such a small parallax also improved the results of his investigations into the motion of the Sun. His particular value, about equal to those of Remus and Horrocks, followed from the proportionality of the planetary diameters and distances.He had thereupon justified this small solar parallax by means of lunar dichotomies.

As in all astronomical matters, Riccioli's *Almagestum Novum* of 1651 contained a complete review of the problem of sizes and distances. Riccioli reported the measurements and opinions of virtually all astronomers since Ptolemy and included the measurements made by himself. Of all the reported measurements of apparent diameters, Riccioli preferred those of Hortensius, which agreed reasonably closely with his own. In his table he drew a line after the pretelescopic values, a line separating, so to speak, the "ancients" from the "moderns." The moderns were divided according to their allegiance, and it is clear that only the Copernicans had contributed anything new to this subject—with the exception, however, of Riccioli himself, whose figures were reported at the end, separated by a line from those of his predecessors.[43]

In order to translate angular diameters into actual ones, Riccioli needed one absolute distance. In principle, he pointed out, the solar distance could be determined in three ways: direct comparison of the solar altitude with the altitude predicted by theory, the eclipse diagram, and the method of lunar dichotomies. The first method was, however, not practical outside the torrid zone because of the problem of refractions at low elevations. The eclipse method was also not suitable because it was hardly possible to avoid an error of half a minute in measuring the apparent diameters of the luminaries and, too, because it was very easy to err by two or three minutes in measuring the width of the shadow.[44] This left only the method of Aristarchus.

Having read Kepler's appeal in his ephemerides for 1619, Riccioli tried the method of lunar dichotomies with the naked eye as well as the telescope. He decided that the best procedure was to pick a month in which the Moon is very close to the ecliptic and to measure two moments of time: first, the moment when the observer begins to think that the Moon is exactly bisected, and second, after the Moon has appeared bisected for some minutes, the moment when he begins to think that the dichotomy has passed. The intermediate time would be the exact moment of dichotomy. The time interval between dichotomy and quadrature (or vice versa when the Moon is waning) was measured, and the angular separation between the Sun and Moon at the moment of dichotomy was calculated from this.[45]

Using this procedure, Riccioli had measured the time interval to be 59 min. 48 sec. on 16 October 1646, and 58 min. 33 sec. on 13 June 1648. From these time intervals followed angular separations between the dichotomies and quadratures of 31′34″ and 30′15″, respectively, leading to angular distances between Sun and Moon of 89°28′26″ and 89°29′45″. Given the lunar distances on those days, Riccioli calculated solar distances of 7,260 e.r. and 7,572 e.r., deciding on a mean of 7,300 e.r.[46] He had also consulted with two lunar experts, Wendelin and Michel Florent van Langren or Langrenus (1600–1675), the cosmographer of King Philip IV of Spain. Wendelin had given him his result of 14,656 e.r., obtained from lunar dichotomies, and Langrenus had stated his solar distance to be 3,420 e.r., but he did not say what the basis for this figure was. Riccioli's solar distance was, thus, conveniently the mean proportional between the values of these two experts.[47]

In preparing his *Almagestum Novum,* Riccioli had also written to other observers for their opinions on this matter. Vincentius Mutus wrote back from Majorca that the method of lunar dichotomy was as uncertain as the method of eclipses. When to the naked eye the Moon still appeared less than bisected to him, the central parts, observed through the telescope, appeared already bisected.[48] This must have been a problem especially for observers who still used a so-called Dutch telescope with a concave ocular. At a typical magnification of 20, such instruments could only show a part of the Moon at one time. Mutus further pointed out that because of the Moon's libration it does not always present the same face to the Earth at the quadratures. Therefore the method could not be standardized by the use of moon maps.[49]

Another correspondent, Giovanni Antonio Rocca, had tried the method with a telescope and had found no sensible time difference between the moments of dichotomy and quadrature. The solar distance, he argued, must therefore be much greater than had been supposed hitherto. According to Rocca, the problem of solar distance and parallax was one of the most important in astronomy, well worth a lifetime's work by any astronomer.[50] In spite of these skeptical comments, Riccioli did not abandon the method of Aristarchus. He advised observers to use a telescope, to observe only the central parts of the Moon, and to choose only dichotomies occurring when the Moon was on or very near the ecliptic.[51]

With a mean solar distance of 7,300 e.r., Riccioli constructed a scheme of sizes and distances (table 14). The distances were, in his Tychonic scheme of the universe, measured from the Earth:[52]

TABLE 14 Riccioli's Scheme of Sizes and Distances

	Apparent Diameter		Distance (e.r.)		Diameter	Volume
Planet	Apogee	Perigee	Apogee	Perigee	(Earth = 1)	(Earth = 1)
Mercury	9″20‴	25″12‴	4,087	10,868	1/4	11/256
Venus	33″30‴	4′ 8″	1,917	12,919	1 15/100	1 1/2
Mars	10″ 6‴	1′32″	2,373	21,005	52/100	14/100
Jupiter	38″18‴	1′ 8″46‴	26,441	47,552	8 4/5[a]	685
Saturn[b]	46″	1′12″	57,743	90,155	20 1/6[a]	891

[a]In calculating the actual diameters of Jupiter and Saturn, Riccioli erred by a factor of 2. Jupiter's actual diameter should be 4 2/5 the Earth's, and Saturn's 10 1/12. The volumes should be 86 and 112.

[b]The measures for Saturn's apparent diameters and actual diameter include the ansae, i.e., rings. The volume, however, is for the planet's body only.

In spite of all its shortcomings, Riccioli's scheme of sizes and distances was very important. The apparent diameters of the planets were based entirely on measurements, and the distances were the largest yet to appear in print. This great expansion of the sphere of Saturn led Riccioli to make the distance of the fixed stars about 200,000 e.r.[53] Thus, even orthodox astronomers were expanding the universe.

Riccioli would have been amiss had he not made his own investigation of atmospheric refraction. As we have seen, the parallax and refraction corrections are of opposite sign, and an error in one will produce errors in the other. Assuming a horizontal solar parallax of about 3′, Tycho had arrived at three separate tables, one for the Sun, one for the Moon, and one for the fixed stars. Riccioli's empirical approach led him more or less to the same results as Tycho, with one difference, however. Riccioli assumed a horizontal solar parallax of 30″ in his investigation, a value, although better than Tycho's, still too large by about a factor of three. As a result, the refraction corrections for the summer were different from those for the winter, and both differed from the corrections to be used in spring and fall. He therefore had not three but nine refraction tables.[54] Had Tycho been more attentive to seasonal differences, he would have found the same phenomenon. When Riccioli's young colleague, Giovanni Domenico Cassini (1625–1712), took up this investigation a few years later and followed Kepler's example in using a theoretical approach, the connection between the magnitude of the solar parallax assumed and the different seasonal refraction tables for different heavenly bodies became one of the main reasons for improvements in the solar parallax (see chap. 11).

In his complete review of the subject, Riccioli listed no fewer than eight different ways of determining the apparent diameter of a planet or fixed star with the telescope. All eight involved comparisons with angular dimensions that were not accurately known themselves. The two methods mentioned by Galileo in his *Sidereus Nuncius*—looking with one eye through the telescope while at the same time looking with the other (naked) eye at an object of known angular diameter, and estimating what portion of the field of view was occupied by the object in question—had not been surpassed in accuracy. Only the measurements of Mercury and Venus during transits presented better opportunities, and even then the angular diameter of the Sun was not known accurately. One could compare a planetary diameter with the separation between two stars close to each other, but how was this separation itself determined accurately? One could compare a planet to the Moon during a conjunction, but what was the angular diameter of the Moon on that particular night? One could estimate what portion of the field of the telescope was occupied by a planet, but determining the field of a telescope was itself not at all an easy procedure.[55]

And to all these vagaries had to be added the poor optical quality of telescopes which hardly ever showed a planet sharply defined. No wonder, then, that the angular diameters were, in retrospect, still much too large and varied in accuracy according to the brightness of the planet. No wonder also that observers continued to measure apparent diameters of fixed stars (although a few began to think of them as dimensionless points of light).[56]

The few new schemes of apparent diameters that we have examined were based on one or two good measurements, such as Gassendi's comparison of the diameters of Mercury and the Sun during the transit, and were completed by comparing the measured planet with other planets at appropriate conjunctions. These comparisons were sometimes performed in more than one stage. Thus, one might compare Venus to Mercury and then compare Jupiter to Venus some time later. Obviously the errors were cumulative. The results varied widely in accuracy from planet to planet and from observer to observer. Only a consistent standard of comparison, equally applicable to all heavenly bodies, could change this.

Moreover, when these measurements or comparisons had been made, there still remained the problem of finding one absolute distance in order to convert these angular measures into linear ones. And here the traditional notion of how far away the Sun ought to be was by no means dead. The Copernican Ismael Boulliau (1605–94), perhaps Europe's most important astronomer during the middle decades of the century, carefully derived a solar apogee distance of $1{,}485^{56}/_{60}$ e.r. by the eclipse diagram in his *Astronomia Philolaica* of 1645,[57] and Gassendi did not stray far from orthodoxy in his *Institutio Astronomica* of 1647.[58] In the printed sources as late as 1655, only Kepler, Wendelin, and Riccioli departed from the traditional figure of the Sun's distance in any significant way. The next two decades, however, brought fundamental changes.

* 11 *
The Micrometer from Huygens to Flamsteed

Although in the course of the seventeenth century the telescope changed astronomical practice profoundly, it did so slowly and in stages. In the first half-century of its existence, after Galileo's dramatic revelations in 1610, the telescope was a qualitative instrument only. Although its potential for precision measurements had been recognized very early, this goal remained elusive (with the isolated exception of Gascoigne's micrometer) until after 1659 when Huygens published a description of a primitive but effective micrometer, possible only for the newer, so-called astronomical telescope. His publication led some of his French colleagues to construct screw micrometers, and this in turn caused the English to "rediscover" and publicize Gascoigne's earlier micrometer. The micrometer had a profound effect on the determination of planetary apparent diameters. Where before astronomers could only estimate the apparent size of a body, and their results varied widely, now they could measure it, that is, they could compare it directly with a known standard, and their results were strongly convergent. The micrometer thus changed these measurements from an "art" to a "science," and by 1675 a consensus was emerging on the apparent sizes of the planets to within a few seconds of arc.

The so-called astronomical telescope, an optical system consisting entirely of convex lenses, had first been described by Kepler in his *Dioptrice* of 1611.[1] His description was, however, a theoretical curiosity without any obvious advantages and a very obvious disadvantage: whereas in the "Dutch" telescope, with its concave ocular, the image is erect, in the simple astronomical telescope it is inverted. It was some time before the advantages of this configuration of lenses were recognized and the instrument began to compete with the "Dutch" telescope. The major advantage of the astronomical telescope was its much larger field of view at equal magnifications.

Because of its small field of view, the full potential of the "Dutch" telescope had been reached very quickly after its invention. By the spring of 1610 Galileo had an instrument that magnified 30 times, and two years later Thomas Harriot had one that magnified no fewer than 50 times.[2] At these magnifications, however, the fields were so small as to make the instruments virtually useless for observing the heavens. Astronomers rarely used "Dutch" telescopes with magnifications of more than about 20.[3]

The astronomical telescope with its larger field broke this bottleneck. Beginning in the 1630s, when the instrument began to come into practice, magnifications began to increase. By the middle of the century the astronomical telescope had replaced the "Dutch" telescope in most serious astronomical work, and magnifications of 50 or more were not unusual.[4]

Another important advantage of the astronomical telescope, slow to be recognized, was that, unlike the "Dutch" telescope, it has a positive focus inside the instrument. This means that an object introduced into the focal plane of this instrument will appear in sharp focus and superimposed on the object under consideration. A common standard of comparison, a ruler, could now be introduced directly into the telescope.

Some time around 1640 William Gascoigne (1612?–44), an astronomer and instrument maker in the north of England, noticed this property of the astronomical telescope by accident, when a spider had spun a web in and around his telescope. Some threads of the web which happened to be in the telescope's focal plane were sharply outlined in the field of the instrument.[5] Gascoigne exploited this optical property by building a micrometer. Writing to the mathematician William Oughtred (1575–1660) in 1641, he described his instrument and reported some of his measurements: on 4 September 1640 Jupiter's apparent diameter was 51″ and Mars's 38″; on 3 January 1641 Mars and Venus were both 25″, " . . . as near as I could discern their limbs by such ill-suited glasses as yet I am worth."[6] Working in a provincial area not well served by instrument makers, Gascoigne could not easily obtain high-quality lenses. In retrospect, the varying quality of his telescopes is reflected in the varying accuracy of his measurements, illustrating that in a telescope equipped with a micrometer the quality of the lenses was now the most important factor limiting the accuracy of the measurements.[7] Gascoigne concentrated his efforts on the apparent diameters of the Moon and Sun, important research topics at that time, and his statement to Oughtred shows that he was fully aware of the importance of his contribution:[8]

> I have shewed some part of these [measurements of the Moon's apparent diameter] to two of my very late acquaintance, both industrious astronomers [Horrocks and Crabtree?], who are not a little taken with them, and the more, because they believed they had tried all possible means, that glasses could afford for the measuring of diameters, yet never attained to any more than guessing, and indeed were so confident, as they would not believe until I let them see a glass to try the moon's diameters by.

Gascoigne died in the battle of Marston Moor on 2 July 1644, and his micrometer remained in private hands, unknown to the scientific community. Not until 15 years after his death were astronomers notified by Christiaan Huygens of a way to turn the astronomical telescope into an instrument for measuring small angles. And not until the late 1660s, when the French had already developed Huygens's idea into the screw micrometer, did the world of science learn about Gascoigne's micrometer.[9]

TABLE 15 Comparative Accuracy of Huygens's Apparent Planetary Diameters at Perigee

Planet	Modern	Hortensius	Horrocks	Riccioli	Huygens
Venus	64″	100″ (56%)	76″ (19%)	248″ (288%)	85″ (33%)
Mars	25″	64″ (156%)	120″ (380%)	92″ (268%)	30″ (20%)
Jupiter	49″	62″ (27%)	60″ (22%)	69″ (41%)	64″ (31%)
Saturn	22″	42″ (91%)	30″–60″	48″ (118%)	30″ (36%)

In 1659 Christiaan Huygens (1629–95) published *Systema Saturnium,* in which he put forth his ring theory to explain Saturn's puzzling appearances.[10] The book also contained Huygens's ideas about the sizes and distances of the planets, as well as his method of measuring their apparent diameters. Huygens introduced tapered strips of copper into the focal plane of his astronomical telescope and determined at what length and thickness a strip just covered the disk of a planet. His telescope was provided with an aperture ring in the same focal plane, so that its field was clearly delimited. Huygens knew the linear diameter of the aperture ring and had determined the angular extent of the field it afforded by timing the passage of a fixed star through this field. He could thus measure the thickness of the copper strip at the appropriate point with a pair of calipers, divide this thickness by the actual diameter of the aperture ring, and multiply this fraction by the angular field of his telescope to obtain the angular diameter of the planet.[11] The standard of comparison was then brought into the telescope itself and precisely determined.

Of course, this "micrometer" could not diminish the glare around bright planets, a feature that had always affected the accuracy of estimates of the apparent diameters of the brighter planets. To cut down this glare, Huygens adopted the practice of coating the eye lens of his telescope with a thin layer of soot from a candle,[12] a practice that had to be exercised with care.[13]

The apparent diameters measured by Huygens in this manner are compared in table 15 with those of Hortensius, Horrocks, and Riccioli, and all are compared with the modern values. The error percentages are given in parentheses in each case. Since Hortensius, Horrocks, and Huygens did not measure Mercury, that planet is not included here.[14]

Several things are apparent from table 15. Not surprisingly, because of its rather large, well-defined disk and lack of excessive brightness, Jupiter was the easiest planet to measure. The smaller and brighter a planet's disk, the more difficult it was to measure its apparent diameter. Thus, Venus's disk was difficult to measure because of its brightness, and Saturn's because of its smallness. Mars was doubly difficult because its disk was bright and small. Mercury, not shown here, was the most difficult planet in all respects. Huygens did very well on the bright planets, partly because of the soot coating on his ocular.

More important, it is apparent from table 15 that the pre-1659 measurements show no sign of progressive improvement. In fact, Hortensius's scheme, the first complete set of telescopic angular diameters, surpassed the later ones in accuracy! It is therefore fair to conclude that before the micrometer the measuring of apparent diameters was much more an art than a science: Hortensius's art was simply better than Riccioli's art.

Furthermore, the lack of a consistent standard of comparison in the earlier measurements is clearly apparent: Hortensius's margin of error varied from 27 percent to 156 percent, Horrocks's from 19 percent (Venus during the transit) to 380 percent, and Riccioli's from 41 percent to 248 percent. In stark contrast, Huygens's results were not only better, they were also much more uniform, varying in error from 20 percent to 36 percent. Clearly even such a primitive micrometer brought tremendous improvement to these measurements.

Like his predecessors, however, Huygens wanted to know the actual sizes of the planets, and for these he needed to know one absolute distance. After examining the traditional methods for determining absolute distances, he was very critical of his predecessors:[15]

> The astronomers try to solve this problem as follows: they first express the distance which separates the Earth from the Sun in terrestrial diameters, and then they draw from this ratio the magnitudes desired. But in estimating that distance from the Earth to the Sun they differ much among one another: which is not surprising, since no tolerable method for measuring that distance has yet been found. For whether they try to discover it by means of eclipses or the dichotomies of the Moon, it can easily be demonstrated that these efforts have to be in vain.

Huygens expressed here what was becoming the common verdict of his generation: the traditional methods of determining the solar distance and parallax were useless. But how, then, was one to proceed? Here Huygens could only fall back on speculations based on harmony.[16]

> That is why it seems to me that only one method remains . . . for forming an idea which is at least plausible of the sizes and distances of all the planets with respect to the Earth. This is to measure with the telescope the apparent diameters of the planets, to investigate next the size of each compared to that of the Sun . . . , and then, having considered them all, to adopt for the size of the Earth with respect to the others that which seems to accord best with the order and disposition of the entire system. In this way, since the ratios of the diameters of the Earth and the other planets to that of the Sun have been determined, and since it is also known how many of its diameters the Sun is distant from us—namely from the angle which its apparent diameter subtends—now the magnitude of the Earth with

> respect to the Sun is also known, and at the same time the distance of the Sun from the Earth as well as from the other planets can be expressed in terrestrial diameters.

Huygens thus followed in the footsteps of Kepler, Remus, Horrocks, and Wendelin. One should look for a proportion and pick a size of the Earth which best agreed with the "order and disposition of the system." It should be noted, however, that whereas these predecessors had been given great leeway by their vague measurements to accommodate planetary diameters to a priori proportions, Huygens's more definite measurements left him little room for such accommodations. Knowing the relative planetary distances, he calculated how large the planets would appear if they were all at the same distance from the Earth as the Sun is. This gave him their proportions to the Sun:[17]

	Sun	:	Planet
Venus	84	:	1
Mars	166	:	1
Jupiter	11	:	2
Saturn	37	:	5

No pattern was readily apparent. Huygens's own measurements showed that the supposition that the planetary sizes were in some proportion to their heliocentric distances was not supported by the observations: Venus came out to be larger than Mars, and Jupiter larger than Saturn. This left him in a quandary. Since the planetary diameters did not uniformly increase with their distances from the Sun, especially Venus and Mars, he could not make the Earth intermediate in size between Venus and Mars. Yet, generally speaking, the outer planets were larger than the inner ones, and Huygens had eliminated all other avenues of finding an absolute distance. He therefore pressed ahead, despite the inconsistencies:[18]

> Nevertheless, in order that the harmony of the entire system can be conserved as much as possible, it appears after all that it is most reasonable to admit that, since the Earth is placed between Mars and Venus with respect to the distances, it also occupies an intermediate place with respect to size. We have said that the diameter of Mars is 1/166th the diameter of the Sun, and that of Venus 1/84th. Taking, then, for the Earth's diameter the mean of these two diameters, we find that it is 1/111th of that of the Sun.

From this Huygens calculated the unheard-of solar distance of 12,543 terrestrial diameters, or 25,086 e.r. Now, Venus's diameter is, in fact, about 1/112th the Sun's, and Mars's diameter 1/204th. Yet, purely by accident, Huygens's value for the Earth's diameter is very close to the figure of 1/109th the Sun's diameter accepted today. As a result, his solar distance was within a few percent of the modern value. Huygens was, of course, not privy to our hindsight, but he was fully aware of the tenuous nature of his argument and conclusion:[19]

> I grant that the calculations rest on a slippery basis, since, to be sure, I have taken the magnitude of the Earth intermediate between those of Mars and Venus on no other ground than that of verisimilitude, and I could thus easily have erred from the truth by some thousands of terrestrial diameters. But even if I have determined those intervals perhaps two times larger or smaller than they are in reality—even three times—it ought nonetheless not to be despised that I have at least determined the measure to that degree [of accuracy], since no other method is at hand in which an error even ten times greater is not to be feared. That, at least, is my opinion.

Huygens had resorted to the harmonic arguments of Kepler without acknowledging his debt to him, preferring, as Remus, Horrocks, and Wendelin had, diameters to volumes in his proportion. Kepler's overarching physical and mathematical construction of the cosmos, a construction in which his own proportion was imbedded, was entirely absent here. Huygens did not even cloak his speculation in astronomical justifications, as Horrocks and Wendelin had done. His speculation therefore appears all the more arbitrary to us. Yet, if it seems unfair to us that he condemned the traditional methods of finding the solar distance while in the same breath replacing them with a method that has a suspicious resemblance to mere guessing, we must realize that Huygens was appealing to a notion of harmony that exercized a powerful influence over him and many of his contemporaries and that to him was wholly "modern" and entirely reasonable. To illustrate this point further, in the dedication of *Systema Saturnium* to Prince Leopold de' Medici (1617–75), Huygens wrote concerning the satellite of Saturn (Titan) recently discovered by him:[20]

> But the longer this neighbor of Saturn was concealed, and the greater the effort needed to draw it down to Earth, the more the discovery is to be rejoiced over. And because the one lacking has now finally completed the collection of wandering stars, and their number being twelve, I almost dare to assert that it will not become greater after this discovery. Certainly, the small ones now exist in equal number to these large and primary ones among which the Earth is to be counted, and both [groups] consist of a number which we consider as perfect, so that this measure can be seen as prescribed by the plan of the highest artisan.

There were now six planets and six satellites, and six was considered a perfect number by the Pythagoreans because the whole is equal to the sum of its divisors: $6 = 1 + 2 + 3$. What are we to make of this statement if we are not willing to admit that, at least at this stage of his career, Huy-

TABLE 16 Apparent Diameter at Mean Distance in Streete's *Astronomia Carolina*

Mercury	10″
Venus	21″
Mars	8″
Jupiter	48″
Saturn	21″

gens believed that intelligent design manifested itself in mathematical harmony, in proportions that existed in nature and that were waiting to be discovered by the student of nature?[21]

Huygens was clearly convinced that the Sun's distance was much greater than the ancients had made it, and while he admitted that he might have erred by as much as a factor of three, he did not rule out the possibility that his successors might discover that his estimate had, in fact, been three times *too small.* The break with the "ancients," that is pretelescopic astronomers, was complete: Huygens did not even bother to discuss the pretelescopic planetary apparent diameters. He was merely improving on the efforts of his contemporaries.

These contemporaries had hardly reached agreement on a solar distance and parallax. The influential Boulliau, for instance, sent him observations of solar eclipses with calculations that incorporated a solar parallax 20 times as large as his own 8.2″.[22] Without any sign of a consensus on solar distance and parallax, and with the traditional methods for determining them now largely discredited, Huygens's speculation was as good a procedure as any.

In this sentiment he was joined by the English astronomer Thomas Streete (1622–89). Streete obtained access to Horrocks's works in the late 1650s, and he made good use of them. In 1661 he published *Astronomia Carolina,* his magnum opus, from which Isaac Newton, John Flamsteed, and Edmond Halley all learned their astronomy. Streete published a list of apparent planetary diameters at mean distance calculated from Gassendi's measurement of Mercury, Horrocks's measurement of Venus, and Huygens's measurement of the superior planets (see table 16).[23]

For determining absolute distances, Streete had tried Aristarchus's method with a telescope, but it appears he did not trust it sufficiently to make it the sole basis for his solar distance and parallax. Resorting to harmonic arguments, he chose Horrocks's proportionality of the planetary diameters and heliocentric distances, using lunar dichotomies only in a supporting role:[24]

> But since our worthy English man Mr. *Jeremy Horrox,* comparing his own observations with others, hath sufficiently proved, that the greatest Parallax of *Mars* in opposition to the Sun, is scarce at all observable, and never amounting fully to 1′, by which and his excellent Telescope observation of *Venus*

> in the Sun, with her apparent diameter at that time, and other good arguments, he determines the Sun's Horizontal Parallax 15″ and no more; which small quantity, agreeing well with the most diligently observed Semidiameters of all the Planets, and being farther confirmed by all our best Telescope observations of the Moon's Dichotomies and otherwise, we accept of, as nearest the truth, and sufficiently exact.

After translating the apparent sizes to real sizes, making Jupiter and Saturn much larger than the inner planets, Streete stated some proportionalities or correlations in the solar system without connecting them to any deeper structure of mathematical or physical astronomy. They were mere statements of the phenomena as he saw them:[25]

> Here it may be well worth our observation, that those Primary Planets that have Secondary or Attendants about them, are of greater Magnitude then those that have none, and
>
> The More Attendants.
> The Greater Orbe.
> The Lesser inequality.
> } Compared according to their number and proportion, are testimonies of the Greater Planet

It was known that Saturn, Jupiter, and the Earth had attendants and that the smaller planets—Mars, Venus, and Mercury—had none. It was also the case in 1661 that Jupiter had more moons than Saturn and was, according to all calculations, also larger than Saturn. The second term is, of course, the proportionality of planetary sizes and distances, expressed here perhaps more as a tendency than as a hard and fast rule. The third term, the smaller the planet's inequality (or, as we would say, eccentricity), the larger its body, was obviously based on the large eccentricities of Mars and Mercury, the smallest planets. We are reminded here of Ptolemy's correlation of decreasing planetary inequality with increasing distance from the central Earth.[26] Because of the popularity of *Astronomia Carolina,* Streete's speculations did not fall on deaf ears.

Streete named his *Astronomia Carolina* after Charles II, the new king of England. The book appeared shortly after the king's coronation, which occurred on 3 May 1661 (on the Gregorian calendar). On that same day, Mercury passed across the Sun. Streete and Huygens, who was in London at that time, forewent the earthly ceremony in order to witness what they considered the more important celestial event.[27] The sky in London was partly cloudy, and the site of their observation, the shop of the telescope maker Richard Reeves on Longacre, was less than ideal. The observation in London was, therefore, a disappointment, although Huygens and Streete did manage to make estimates of Mercury's apparent diameter and position.[28]

In Gdansk, Hevelius had more luck. He had an unobstructed view of the transit for quite some time and was able to make a number of measure-

TABLE 17 Hevelius's Apparent Diameters

Planet	Apparent Diameter		
	Min.	Mean	Max.
Mercury	4″ 4‴	6″ 3‴	11″ 48‴
Venus	9″ 34‴	16″ 46‴	1′ 5″ 58‴
Mars	2″ 46‴	5″ 2‴	20″ 50‴
Jupiter	14″ 36‴	18″ 2‴	24″ 22‴
Saturn	10″ 34‴	12″ 20‴	14″ 48‴

ments. While in London, Huygens had obtained a copy of Horrocks's still unpublished *Venus in Sole Visa,* which he now sent on to Hevelius.[29] Hevelius printed it as an appendix to his own *Mercurius in Sole Visus,* which appeared in 1662.

Needless to say, all observers of this transit used telescopic projection methods, but Streete and Hevelius did not yet have micrometers. While Streete estimated Mercury's apparent diameter to be at most 1⁄100th of the Sun's,[30] Hevelius judged it to be no more than 1⁄160th. Taking the Sun's apparent diameter to be 31′28″ on that day, he calculated Mercury's apparent diameter to be 11.8″; this was the planet's angular diameter *at perigee.* Hevelius now calculated it to be 6.05″ at mean distance and 4.1″ at apogee.[31] He thus made Mercury considerably smaller than had his late colleague and correspondent Gassendi. Hevelius also made the apparent diameters of all other planets much smaller than anyone else had made them (see table 17).[32]

This was the first time (in retrospect) that anyone had erred on the small side in measuring apparent diameters. Hevelius had determined these apparent diameters by estimating what portion of the field of his telescope was taken up by the disk of a planet. He had, however, used extremely small apertures: for observations of very bright bodies such as Mercury or fixed stars, he used an aperture the size of a pea, that is, about 1⁄4 inch![33] We now know that such very small apertures cause noticeable diffraction effects, so that one does not observe the actual disk of the star or planet so much as a spurious disk. Although in 1660 astronomers could hardly have known this explanation, Hevelius's colleagues did realize that very small apertures gave spurious results. His measurement of Mercury during the transit was, therefore, the only one considered of any use by other astronomers.[34] Hevelius made the solar parallax 40″ and the Sun's distance 5,000 e.r. without explaining how he had arrived at those values.[35]

Within a few years the primitive micrometer described by Huygens in *Systema Saturnium* in 1659 led to the development of the screw micrometer in France and its rediscovery in England. The measurement of apparent diameters now rapidly became a routine procedure with consistent results.[36] By 1666 Jean Picard (1620–83) in Paris was launched on a systematic program to measure the angular diameters of the Sun, Moon,

TABLE 18 Flamsteed's Apparent Diameters

Planet		Flamsteed	Modern	% Error
Venus	min.	10″ 21‴	9.9″	4.5
	max.	1′ 12″ 16‴	1′ 4.0″	12.9
Mars	min.	4″ 48‴	3.5″	37.1
	max.	34″ 24‴	25.1″	37.1
Jupiter	min.	34″ 16‴	30.5″	12.3
	max.	53″ 50‴	49.8″	8.1
Saturn	min.	19″ 10‴	14.7″	30.4
	max.	26″ 36‴	20.5″	29.8

N.B.: Flamsteed's actual measurements were to the nearest half or third of a second. The thirds given here are the results of his calculations.

and planets,[37] and by 1671 John Flamsteed (1646–1719) was doing the same in Derby.[38] Picard did not systematize his planetary measurements, but Flamsteed did. In a little manuscript tract prepared in 1673, he calculated the apparent diameters of all planets, except Mercury, at apogee and perigee from his own micrometric measurements. They are given in table 18 with the modern values for comparison.[39]

As can be seen, Flamsteed's values were an improvement over Huygens's values (see p. 120). Picard's measurements may, in fact, have been even a little better.[40] The measurement of apparent diameters had been reduced to a routine procedure with reproducible results by the micrometer. Over the next few decades most observatories were equipped with micrometers, so that by the turn of the eighteenth century planetary diameters could be measured easily and accurately by all well-equipped observers. The new consensus on the apparent diameters of planets emerged quickly after 1670, and the values were within reasonable margins of the modern ones. The errors were uniform, and they were also positive. This indicates that they are to be ascribed more to the imperfect optics of the telescopes than the inaccuracies of the micrometers. The consensus has been a lasting one. From the time of Flamsteed and Picard, there have been no dramatic changes in the measurements of apparent planetary diameters. The practice has become progressively more refined as telescopes and micrometers have been improved, and the consensus has come within progressively narrower bounds.

The same optical property that made the micrometer possible in the astronomical telescope also allowed the installation of cross hairs in the optical plane. This insured accurate alignment of the telescope, elusive until then, and thus made possible the use of telescopic sights on measuring arcs. While the micrometer allowed the measurement of small angular distances within the field of the telescope, the telescopic sight revolutionized the measurement of larger angles. The vaunted accuracy of Tycho Brahe's observations was, therefore, definitively surpassed in the last quarter of the seventeenth century. Within the narrower error margins, the effect of solar parallax and refraction on solar theory now became a major problem in predictive astronomy. The growing consensus on apparent planetary diameters was followed by a lively debate on absolute distances, the most authoritative values for which resulted from attempts to improve solar theory.

12
Cassini, Flamsteed, and the New Measure

It is commonly supposed by historians and astronomers that the first successful determination of any parallax of a heavenly body other than the Moon resulted from simultaneous measurements of Mars's positions in France and Cayenne in 1672. According to the usual accounts, Giovanni Domenico Cassini (1625–1712) derived a parallax of Mars of about 25″ from the observations, which made the solar parallax 9.5″ and the solar distance about 87,000,000 of our miles.[1]

Although the historical record shows clearly that Cassini did, indeed, arrive at these figures, an analysis of his method shows that he could have extracted a wide variety of parallaxes of Mars from the observations. Mars's parallax was swamped by the error margins of the measurements, and the literature usually grants Cassini's results great accuracy where none existed. The nearness of Cassini's parallaxes of Mars and the Sun to our present-day values for these measures must not be taken as an indication of their accuracy. But if Cassini had wide latitude in determining Mars's parallax, why did he choose 25″ and a solar parallax of 9.5″? The answer to this question is found in his attempts to eliminate residual errors in solar theory (affecting the theories of all the planets) by finding a better combination of parallax and refraction corrections.

With telescopes fitted with screw micrometers and measuring arcs equipped with telescopic sights, the accuracy of astronomical measurements improved dramatically. Small quantities which had hitherto been indetectable could now be measured; discrepancies which hitherto had been hidden now became glaring. The problems of atmospheric refraction and solar parallax, and the relationship between them, now became crucially important. In the Sun King's newly founded Académie Royale des Sciences in Paris, they were the subjects of constant discussion and numerous proposals.[2] Little progress was made, however, until after the arrival in Paris of Giovanni Domenico Cassini in 1669.

Cassini was appointed professor of astronomy at the University of Bologna in 1650 at the early age of 25, and he turned his attention to the motion of the Sun which was becoming a pressing problem. At that time Bologna boasted what surely must have been the world's largest gnomon. High in the southern face of the cathedral of San Petronio was a small aperture that cast an image of the Sun on the cathedral's floor. Near noon this image crossed a line at an angle of 9° to the meridian: the pillars made a true north-south line impossible. Turning the cathedral into a giant pinhole camera had been the idea of cosmographer Egnatio Danti (1536–86). Alterations to the cathedral had made Danti's meridian unusable; therefore Cassini constructed a new and larger gnomon with a meridian that was, this time, exactly north-south. This gnomon can be seen in the cathedral today. In 1653 Cassini could proceed with his measurements.[3]

With the new gnomon, Cassini measured the Sun's meridian altitudes throughout the year and corrected them for refraction and parallax as best he could. From the refined measurements, he knew the Sun's maximum and minimum declinations, occurring at the solstices; half the difference of these was the obliquity of the ecliptic, a fundamental parameter in astronomy. But how accurate was this determination? When Cassini took half the sum of the declinations, which gave him the Sun's declination at the equinoxes and thus located the celestial equator in Bologna, he found that it was not exactly 90° removed from the celestial pole determined by measurements of the circumpolar stars. The difference of 2¼′ was significant.[4]

After a thorough examination of all the possible sources of error, Cassini concluded that the discrepancy was from faulty corrections for refraction and parallax in his solar elevations. He had used the refraction tables of Tycho Brahe and Riccioli, his older colleague in Bologna, to correct his measurements downward, and Kepler's solar parallaxes to correct them upward.[5] If this combination had caused the discrepancy, then what other combination would reconcile the two sets of measurements?

It appears that Cassini did not think highly of the traditional methods used to determine the solar parallax, for he concentrated his attention on the problem of refraction, familiarizing himself with all its aspects and performing a number of experiments. He was the first to use Snell's law of refraction in these studies. Since, however, there were two corrections, one for refraction and one for parallax, an ambiguity remained. As he related it later:[6]

> Two hypotheses were proposed which at the Sun's meridian altitudes in the climate of Europe caused more or less the same effect, so that there was no sufficiently certain way clearly to distinguish one hypothesis from the other. One supposed that the parallax of the Sun was insensible or less than 12″; and in this hypothesis the refractions were invariable throughout the entire year. The other supposed, like Kepler, a horizontal solar parallax of 1′, and this supposition obliged one to vary the refraction throughout the year in proportion to the variation of the Sun's declination.

The second hypothesis was, in fact, the one found in Riccioli's *Almagestum Novum,* except that Riccioli made the Sun's horizontal parallax 28″. We may assume that upon his arrival in Bologna in 1648, under the tutelage of Riccioli and his colleague Francesco Maria Grimaldi (1613–

63), Cassini had begun with this approach. As we have seen, however, Riccioli's ideas led to separate tables for winter, spring or fall, and summer. The first hypothesis led to a single table, which was more elegant, but it forced one to accept a solar distance that was extremely large.

In his *Specimen Observationum Bononiensum* of 1656, Cassini preferred the first hypothesis, but in his new solar tables published in 1662 in *Cornelius Malvasia's Novissimae Motuum Solis Ephemerides* he used the second.[7] We may speculate that as an astronomer actively engaged in research on this subject he liked the first option for its simplicity, while as a product of his time he found it difficult conceptually to accept a solar distance of at least 17,000 e.r.

When he arrived in Paris in 1669 Cassini found opinions there equally divided. Huygens, the most prestigious member of the Academy in its early years, was of the opinion that in finding refraction corrections empirically by measuring the Sun one "should suppose that the Sun has no sensible parallax at all, as is well known from experience."[8] Picard, on the other hand, was more hesitant. When he found in the spring of 1669 that the Sun suffered more refraction then than it had the previous winter, he stated:[9]

> There is reason to suspect that this is due to the parallax of the Sun. But [to determine] that, a long series of observations made with large instruments in a proper location is needed.

The Academy was housed in the Bibliothèque du Roi, and astronomical observations were made in its garden. Picard pointed out that this site was less than ideal for observations and that the instruments available left something to be desired. He looked forward to the completion of the new observatory which was under construction.[10] Later that year he reported that his observations, when compared to tables and ephemerides, showed clearly that all tables of the Sun's motion were defective, and that the program of taking the Sun's meridian altitudes with large instruments should be continued with particular care. Moreover, a start should be made on preparing refraction tables designed specifically for Paris, "according to the different seasons and even according to the different changes in weather."[11]

When Cassini arrived in Paris, the effect of solar parallax and refraction on solar theory (and therefore planetary theory as well) had thus become central in the program to reform astronomical tables in order to bring them in line with the new accuracy afforded by the telescopic sight and micrometer. Cassini may already have been convinced that a single refraction table with corrections continuing right up to the zenith was the correct approach. This meant that the horizontal solar parallax was at most 12″. The combination produced an obliquity of the ecliptic of 23°29′. Tycho's refraction tables, in which the corrections for the Sun stopped at 45°, led to an obliquity of the ecliptic of 23°31½′.[12] Since the various combinations of refractions and parallaxes had been composed to reconcile the sun's solstice elevations with the observations of the circumpolar stars, the issue could not be decided by means of observations—at least, not observations made at the European latitudes.

What would happen, however, if one made these same observations in the torrid zone near the equator? At a spot at or near the equator, the noonday sun is always 60° or more above the horizon. This meant that in a combination of a solar refraction correction of zero above 45° and a parallax correction based on a horizontal parallax of 3′, such as Tycho had used, the parallax corrections would be the only ones in the tropics. Conversely, in a combination of refractions continuing up to the zenith and very small parallaxes based on a horizontal parallax of 12″, such as Cassini now proposed, the refraction corrections would always be greater than the opposite parallax corrections. According to Tycho, therefore, the Sun's real meridian altitude in the tropics would always be greater than the observed one, whereas according to Cassini the opposite would always be the case.[13]

Moreover, whereas at the European latitudes the Sun is always seen in the south, at or near the equator it is seen in the northern sky in the (European) summer and the southern sky in winter. This meant that in the tropics Tycho's net parallax corrections would be subtracted from the measured arc between the June and December solstices, while Cassini's net refraction corrections would be added to the arc. The two combinations therefore had to produce different obliquities of the ecliptic in the tropics, although at European latitudes they produced the same. Whichever combination produced the same obliquity at both latitudes would be the correct one.

An expedition to the tropics was thus called for, and the French academy had the resources, or rather the cooperation of the government, to undertake such expeditions. In 1670 Picard had measured the length of a degree in northern France with unprecedented accuracy.[14] In 1671 he went to Hveen to measure the longitude and latitude of Tycho's observatory with respect to Paris, and to check some of Tycho's measurements in order to see if perhaps there were systematic errors in Tycho's observations.[15] As early as 1667 there had been discussion in the Academy of an astronomical expedition to Madagascar, but the need for such an expedition had not been pressing, and the proposed astronomical program had been poorly conceived.[16]

There was now another and more urgent reason for an astronomical expedition. All astronomers were familiar with Tycho's atttempted measurements of the parallax of Mars. Kepler's verdict had been that Mars's diurnal parallax was less than the error margin of Tycho's instruments. Now, however, the revolution in instrumentation had greatly reduced that error margin. Where Kepler put Mars's parallax at opposition at 2′ at most, from which a horizontal solar parallax of at most 1′ followed, the margin of error of one of the new measuring arcs with telescopic sights was thought to be at most some seconds of arc. Perhaps, therefore, the parallax of Mars at opposition could now be measured. Mars's position measured at the same time at Paris and at a point of known coordinates far from Paris would, it was hoped, reveal a parallactic displacement that had now become detectable with the new instruments. Any site in the tropics would be far enough from Paris for these measurements. The most favorable occasions for such measurements were oppositions when Mars was at its perihelion, when Mars's parallax is fully 2½ times as great as the Sun's, and such a rare opportunity was to present itself in 1672. There was thus every incentive to press forward with the plan for an expedition to the tropics, and Cassini wasted no time.[17]

The island of Cayenne, in the mouth of a river by the same name, on the coast of South America in what is today French Guiana, had been settled by the French earlier in the century. Now Jean Baptiste Colbert (1619–83), Louis XIV's minister and the organizer of the Academy, was turning his attention to this area in the hope of establishing a profitable colony there.[18] Ships sailed back and forth between Cayenne and the mother country regularly, so that travel would present no problems. Cayenne's location, 5° north of the equator, was ideal for the proposed observations, and its climate, although difficult for Europeans, was not forbidding. For various reasons an expedition to Cayenne had been planned for some time.[19] Now, under Cassini's urging, it became a reality. The young astronomer Jean Richer (d. 1696) and an assistant, a certain Mr. Meurisse, left La Rochelle on 8 February 1672, carrying with them several sizeable measuring arcs with telescopic sights, but apparently no micrometer.[20]

Cassini felt that the determination of the obliquity of the ecliptic, and the light this would shed on refraction and parallax, was alone worth the expedition. Needless to say, however, besides these and routine geographical observations, Richer was asked to make other astronomical observations that could be made in the tropics. His charge from the Academy was to make the following observations:[21]

1. On the true obliquity of the ecliptic.
2. On the moments when the equinoxes arrive, or, which is the same thing, how much more time the Sun remains in the northern signs than in the southern ones.
3. On the parallaxes of the Sun, of Venus, and of Mars; the latter of these planets will be at its greatest proximity to the Earth in the months of August and September of 1672, which only occurs very rarely.
4. On the motions and parallax of the Moon which are not yet well known.

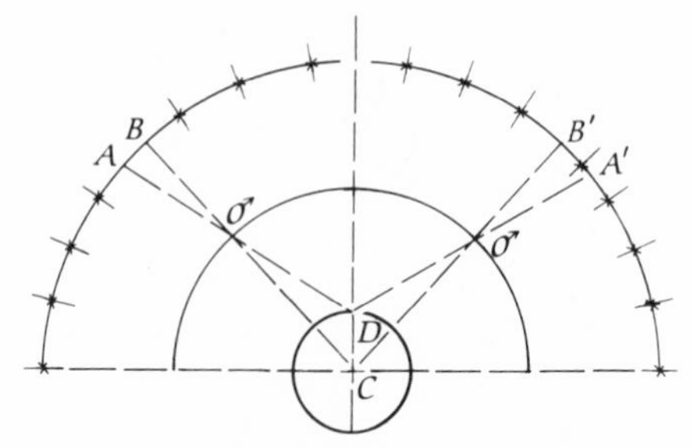

FIG. 12. Diurnal parallax of Mars

5. On the motions of Mercury which is seen only rarely in Europe.
6. On the declination, right ascension, and magnitude of the southern fixed stars which do not appear above the horizon in Paris.

For determining the parallaxes of the Sun, Venus, and Mars, as well as the differences in longitude between Cayenne and Paris, the astronomers of the Academy would make observations of the same phenomena in France.

The observations in connection with the parallax of the Sun and Venus offered little hope of success, but there were very high expectations of the simultaneous observations of Mars. One did not have to go to Cayenne, however, in order to attempt to measure the parallax of Mars. Provided that Mars's parallax was now actually within the grasp of these new instruments, Tycho's method of measuring the planet's positions at the same location four hours before and after its meridian transit would do just as well, as illustrated in figure 12. The observer at *D* would measure Mars's positions at *A* and *A′*, whereas with respect to the Earth's center, *C*, Mars's position had, in fact, changed from *B* to *B′*. Once the appropriate adjustments had been made for Mars's (retrograde) motion during the time interval, the net change in position was simply the sum of the two parallaxes *AB* and *A′B′*. This method would be used by Cassini and his colleagues in France.[22]

* * *

Across the Channel, the young astronomer John Flamsteed (1646–1719) was making similar preparations. Flamsteed, largely self-educated in astronomy, took his notions on solar parallax mostly from Streete, Horrocks, and Riccioli, and he defended Horrocks from Hevelius's charge that he had made this parallax much too small.[23] Flamsteed had read Horrocks's long demonstration of the uselessness of the eclipse method for determining solar parallax, but Horrocks had not definitively rejected Aristarchus's method. Flamsteed had studied Riccioli's discussion of lunar dichotomy in the *Almagestum Novum,* and he agreed with Riccioli that this procedure, if used with care, could give good results. As late as May 1671 he was still trying to determine the solar distance by Aristarchus's venerable method.[24] Indeed, until October 1672 he accepted Riccioli's value of 28″ or 30″ for the solar parallax.[25]

His thoughts on refraction were equally conventional. Since he had no accurate measuring arc, he was not able to study this phenomenon in detail, and he actually suspected, on the basis of some observations by the mathematician Edward Wright (1558–1615), that "in inland places" the refraction suffered by the Sun's light was zero or insensible at elevations of 20° or more.[26]

Flamsteed did not obtain a good telescope until 1670. Not until the autumn of 1671 did he have the instrument, equipped with a micrometer, properly installed for observations at his father's house in Derby.[27] He now concentrated his efforts on micrometrical measurements, such as the apparent diameters of the Sun, Moon, and planets, as well as the separation of any two celestial objects that he could bring into the field of his telescope simultaneously. Thus, he measured the separations between planets and fixed stars during appulses. When the news of the French expedition to Cayenne reached him, he too sought a way to profit from the upcoming favorable opposition of Mars. Tycho's method, referred to in so many books, was obvious and fairly simple. The problem for Flamsteed was to make the measurements with sufficient accuracy.

Flamsteed was convinced that his micrometer measurements of separations during appulses were extremely accurate. In April 1672 he wrote to Henry Oldenburg (1618?–77), secretary of the Royal Society:[28]

> My last trialls persuade mee so much of ye accuracy of this method of observeing which I use at present that I doubt not but if I had once ye opportunity of observeing any transit of Mars in or near his Achronicall appearance [i.e., opposition] by any, tho unknowne, fixed star; I could determine something concerneing his & consequently ye earths distances a Sole [i.e., from the Sun] & parallaxes; better than as yet any one has.

Shortly afterward Flamsteed discovered that at the end of September and the beginning of October of that year, although Mars would already be a little past its favorable opposition, it would pass very close to three little stars in Aquarius. The young astronomer was now even more optimistic:[29]

> . . . I find yt being very carefull my instruments will show 5 seconds, so yt I no longer doubt to determine his parallax if God permit mee health & a cleare sky[.]

Flamsteed now composed an ephemeris of Mars's motion from 25 September to 13 October 1672 (n.s.), during which period the planet would pass by the little stars. Oldenburg printed this in the *Philosophical Transactions* in August.[30] Flamsteed advised his colleagues that according to

Kepler's solar parallax, Mars's parallax would be between 2′ and 2½′, and according to Streete's and Horrocks's solar parallax of 15″, Mars's parallax would be from 31″ to 36″. Even the latter values were "sufficiently large & sensible" quantities to be measured with a good micrometer.[31] As for his own expectations:[32]

> I expect them lesse then Keplers, larger then Streets but findeing how much I have failed in my conjectures formerly I am resolved to let my observations lead mee, not with others to wrest them.

So, Mars's favorable opposition was attended in England, France, and Cayenne. God gave Flamsteed health and a clear sky, but his father sent him on a business trip. Although he was able to spend some time with Richard Townley (fl. 1660–1705), the virtuoso who had inherited and improved Gascoigne's micrometer and made it known to the world, he was unable to observe Mars twice on any night while on this trip. On his return to Derby he was able, finally, on the night of 6 October 1672 (n.s.) to measure Mars's distance from two of the stars several times over a period of 6 hr. 10 min.[33] His parallax calculations were all made from this single set of measurements.

From his measurements, Flamsteed found that during this period of 6 hr. 10 min. Mars had moved 37½″ retrograde in longitude, and 1′19½″ downward in latitude. In a lecture delivered nine years later, he showed how he found the planet's horizontal parallax from this. From the tables he ascertained that in that interval Mars had actually moved 20″ retrograde in longitude and 1′13½″ downward in latitude. Thus, the total parallactic displacement, that is, the sum of the two parallaxes *AB* and *A′B′* in figure 12, was 17½″ in longitude and 6″ in latitude. From this he had to find, by an involved geometrical procedure, how much Mars's parallax was at the times of observations, and from either one of these Mars's horizontal parallax could be calculated.[34]

Flamsteed's first pronouncement on the results came three weeks after the measurements, on 26 November 1672 (n.s.), when he wrote to Oldenburg that Mars's parallax "was very small certeinely not 30 seconds. So yt I beleive the Suns is not much more then 10″."[35] Three months later in February 1673, however, he had revised those figures. He now believed that in September observations Mars's horizontal parallax was at most 15″, which made the Sun's parallax no more than 7″ and its distance at least 29,000 e.r.[36] When in July 1673 he wrote an open letter to Cassini published in the *Philosophical Transactions,* he had again altered his figures. He now believed that,[37]

> the parallax of Mars at opposition and perigee is never greater than 25″, from which it follows that that of the Sun is at most 10″ and its distance 21,000 terrestrial radii.

When he received Cassini's reply, Flamsteed was flattered to discover that the great astronomer had arrived at exactly the same conclusion concerning the parallax of Mars:[38]

> As for the parallax of Mars, it is remarkable how much trouble we took in defining it. For I discovered from very frequent observations last year and especially in the month of September, by comparison with the fixed stars revolving in almost the same parallel, from the variation of the right ascension between [evening] and morning, that the parallax of Mars at that time was 25 seconds, and considering the ratio between the distances of Mars and of the mean Sun from the Earth, I defined the mean solar distance as 22 thousand earth-radii.

Such agreement could not be fortuitous, and in the absence of either man's publication of the actual measurements and the calculations based on them, it was not easy to criticize these results.

* * *

Cassini and the young Danish astronomer Ole Roemer (1644–1710), who had come back from Denmark with Picard in 1671, had made observations at Paris until Cassini had to leave the city at the end of September 1672. Roemer remained in the Paris observatory to continue the observations, while Cassini made observations where and when he could on his trip. In the meantime, Picard was in Anjou and observed Mars there during the favorable opposition.[39]

In attempting to measure the diurnal parallactic displacement of Mars, Cassini and Roemer did not use a micrometer as Flamsteed had. Instead, they used a pendulum clock to measure difference in right ascension between Mars and the neighboring stars. Thus, measured at the same hour on the 8th and 9th of September 1672, Mars's right ascension showed a difference of 67½ seconds of time. The difference between the 9th and the 10th was 66¾ sec. From this Cassini could find Mars's actual motion in right ascension during any smaller period during those days. On the 9th they found that the difference in right ascension between Mars and a nearby fixed star, measured at 8:36 in the evening and then again at 3:56 in the morning, had changed by 21½ sec. According to Mars's daily motion in right ascension, however, the planet had actually moved 19¾ sec. in right ascension over this period of 7 hr. 20 min. This left 1¾ sec. as the apparent motion in right ascension caused by Mars's parallax. Knowing the time of Mars's meridian passage, its declination at that time, and the latitude of Paris, Cassini could now calculate the planet's horizontal parallax. It came out to be 24¾″. Likewise, from measurements made on two other occasions in September, Cassini found parallaxes between 24″ and 27″.[40]

This method, which produced time differences of 1 or 2 seconds attributable to Mars's parallax, obviously was not very accurate. An error in time of even ¼ sec. would result in an error in Mars's parallax of about 4″. When he learned what method Cassini had used, Flamsteed was quick to point out its deficiencies.[41]

> . . . Signiour Cassini pretends to have found [the distance of the Sun] by some observations made about the same time [i.e., September 1672] on the same planet [Mars]. which yet I am confident by his method of observeing (yt is by counteing ye time interlapsed betwixt ye transit of ye planet, & a near star, of ye same declination with it, over a [thread] placd in an hour-circle in his telescope, & converteing it into partes of ye Equator) hee could never discover it, as, if I might have a sight of his observations, perhaps I might easily prove.

And well he might have. In the above cited example of 9 September 1672, Mars's motion in declination over the period of 7 hr. 20 min. accounts for

$$\frac{67\frac{1}{2} + 66\frac{3}{4}}{48} \times 7\frac{1}{3} \text{ sec.},$$

or 20½ sec., not 19¾ sec. Thus, only 1 sec. was attributable to the parallactic difference, and Mars's horizontal parallax should be 14″, not 24¾″.

Cassini himself was not unaware of these problems. On 17 September 1672, his measurements resulted in a horizontal parallax of Mars of 27⅓″. Since 9 September, however, Mars had steadily moved away from the Earth, and therefore its parallax on the 17th should have been smaller, not greater, than the 24¾″ Cassini found on the 9th.[42] Cassini wrote:[43]

> . . . since the differences which we found between the apparent and true variations were only of one or two seconds of time, a large number of observations which most often gave nearly the same result was needed in order to be persuaded that that difference came from the parallactic and not from some error in the observations which are, after all, subject to similar differences and sometimes even larger ones. From which it happened some times that no difference was found between the apparent and true hourly movements, and some times a small difference contrary to the effect of parallax was found. We adhered to what we found most often and by the most select observations.

Neither Cassini nor Flamsteed, however, published the details of their measurements and calculations for quite some time. Cassini, moreover, had used other methods as well. After all, Richer had been sent to Cayenne to make simultaneous observations of Mars. His task had been to observe at least two solstices, and as 1673 progressed his return to France was anxiously awaited. In his history of the Academy, Bernard Le Bovier de Fontenelle (1657–1757) wrote, half a century later:[44]

> The return of Mr. Richer was awaited as one would await the decree of a judge who was to pronounce on the important difficulties which divided the astronomers. One may say that astronomy was in suspense when Mr. Richer arrived from Cayenne toward the end of that year.

Richer returned in August 1673 and the records of his observations quickly showed that the expedition had been entirely successful. The question about the relationship between refraction and solar parallax, which had been the main reason for the expedition, received a definitive answer. Richer's raw measurements were as follows: the Sun crossed the meridian 71°27′40″ above the northern horizon at the June solstice, and 61°35′16″ above the southern horizon at the December solstice. Using Tycho's combination of refraction and parallax corrections, parallax was 55″ (+) at 71½° and 1′28″ (+) at 61½°. No refraction corrections were necessary. The true solstice elevations, according to this combination would thus be 71°28′35″ and 61°36′44″. According to the other combination, surely favored by Cassini by 1673, at 71½° there was a refraction of 20″ (−) and a parallax of 3″ (+), and at 61½° refraction was 32″ (−) and parallax 4″ (+). This made the true solstice elevations at Cayenne 71°27′23″ and 61°34′48″.[45]

Using Tycho's corrections, the observations at Cayenne resulted in a distance between the tropics of 46°54′41″, and therefore an obliquity of the ecliptic of 23°27′20½″. From his own solstice observations at Hveen, corrected by the same tables of refractions and parallaxes, Tycho had found an obliquity of 23°31′30″. Cassini's combination of corrections produced from those same measurements at Cayenne an obliquity of the ecliptic of 23°28′54½″, while his observations with the gnomon of San Petronio, corrected by his refractions and parallaxes, produced an obliquity of 23°29′. Tycho's corrections thus yielded values for the obliquity of the ecliptic near the equator 4′ smaller than in the northern latitudes, while Cassini's corrections produced the same value at both latitudes with very good agreement.[46] Cassini's growing belief that sensible refractions continued to the zenith and that solar parallax was very small, probably not more than 12″, was proved correct in no uncertain terms. Henceforth a single table of refraction corrections for all seasons and for all heavenly bodies would suffice, and there were now excellent astronomical reasons for supposing that the Sun's horizontal parallax was very small.

But how successful had Richer's expedition been for the actual determination of parallaxes? Richer had been instructed to measure the meridian altitude of the Sun as often as possible, and to do the same for bright stars at comparable elevations. Observers in Paris were to do the same. The difference in altitude between the Sun and a star measured at Cayenne, when compared to the same difference measured in Paris on that same day, would reveal any parallactic displacement of the Sun attributable to the difference in latitude between Cayenne and Paris. The larger the number of measurements that could be compared, the greater the chance of detecting a systematic difference. Since, however, the accuracy of modest measuring arcs with telescopic sights such as Richer had available was of the order of 15″, this method was not sufficiently accurate to produce the desired result. Similar measurements of Venus's positions were too difficult to produce useful results.[47] Richer's measurements of Mars held out more hope, however.

Cassini and his colleagues hoped for simultaneous observations in France and Cayenne of Mars's close approach to one of the little stars. If Mars eclipsed a fixed star and this event occurred at exactly the same time in both places, then there was no sensible parallax. If, however, in Paris Mars was observed to eclipse a star while at Cayenne it was observed to pass by it at a certain distance, or vice versa, then that distance was the parallax due to the separation between Paris and Cayenne.[48] In fact, although it appears that Mars did indeed eclipse one of the little stars used by Flamsteed, this event was not observed in either place.[49] No matter what measurements were to be compared, however, it was essential to know the latitude and longitude difference between Cayenne and Paris. Cayenne's latitude was easily determined, but its longitude with respect to Paris was more difficult to find. Observations of the eclipses of Jupiter's satellites and eclipses of the Moon fixed the time difference somewhere between 3 hr. 27 min. and 3 hr. 42 min. Cassini used 3 hr. 39 min. in his calculations.[50]

Like Flamsteed, the observers in Cayenne and France measured Mars's positions with respect to the three small stars in Aquarius near which the planet passed in September and October. But whereas Flamsteed had measured the actual angular separations between Mars and these stars and had then calculated the longitude and latitude components, the French observers measured the difference in right ascension and declination, the former with pendulum clocks and the latter with measuring arcs.[51]

In his calculations of the declination differences, as he presented them later, Cassini chose only three sets of measurements, those made in Paris and Cayenne on 5, 9, and 24 September. From these he used only the declination differences between Mars and the first of the three little stars to cross the meridian. Mars's daily variations in declination were found from measurements of the planet's meridian altitudes, and this allowed Cassini to make adjustments for the time difference between Cayenne and Paris in the declination differences measured at Cayenne. In other words, he referred the measurements made in Cayenne to the meridian of Paris. On 5 September the difference in declination between Mars and the star was 32′22″ in Cayenne and 32′10″ in Paris at the same moment. The parallactic difference was thus 12″. (Note that this difference must have been equal to or less than the accuracy of the measuring arcs with which the declinations were measured.) On the 9th the parallactic difference was 13″, and on the 24th it was 17″.[52] Again, however, on this last day Mars was further from the Earth than it had been on the two previous dates, and therefore the parallax should have been smaller. Cassini commented:[53]

> Thus that augmentation must be attributed to an imperceptible error in the observations, which [error] is most certainly divided equally between the second and third, making the difference 15″ at a time intermediate between the 9th and the 24th of September, such as between the 16th and 17th of the same month.

Cassini thus boldly concluded from only three observations, chosen carefully from a much larger body of similar measurements, that on 16 and 17 September Mars's parallax between the latitudes of Cayenne and Paris was 15″. From this, Mars's horizontal parallax was 25⅓″.[54]

For calculating Mars's parallax from differences in right ascension between Mars and a fixed star, only the observations made at various places on 1 October 1672 were used. At Brion in Anjou, Picard found at 7 P.M. that Mars's western limb trailed the middle one of the three little stars by 4 sec. and at 2:30 A.M. that its eastern limb preceded that same star by 6 sec., while it took Mars's disk 1⅔ sec. to cross his hour line. Over an interval of 7½ hrs., therefore, Mars's right ascension had changed 11⅔ sec. of time. Mars's actual motion in right ascension over that time was calculated by Cassini to be 8⅛ sec. of time, so that the parallactic difference was 3½ sec., or 52″ of arc.[55]

On that same evening, the western limb of Mars passed the meridian line at Cayenne 7 sec. before the middle one of the three stars. Mars's meridian transit occurred at 10:25 P.M. Now Cayenne was 3 hr. 39 min. west of Paris, and Brion was 11 min. west of Paris. The time difference between Cayenne and Brion was, therefore, 3 hr. 28 min., and Richer's meridian observation of Mars occurred at 1:53 A.M. Brion time, that is, only 37 min. before Picard made his second observation. Allowing for Mars's motion during those 37 min., it was found that at Cayenne Mars preceded the star by exactly the same amount as it did at that moment in Brion. Hence no parallax resulted from this comparison.[56] Roemer's observations in Paris that night consisted of estimates without a micrometer of the distances between Mars and the three stars. These observations compared with those of Picard in Brion supported the parallax favored by Cassini.[57]

Therefore, when the work of the observers in Cayenne and various parts of France had been analyzed, Cassini could find few observations that yielded useful results. The right ascension differences measured at the same places but different times produced three sets of observations in Paris which gave a horizontal parallax of Mars from 24″ to 27″, and one set in Brion which gave a parallax of 52″. Right ascension differences measured at different places but at the same times, produced one comparison between Cayenne and Brion that gave no parallax. And one comparison between Brion and Paris at different times produced a result that tended to support a parallax of 25″. Among all the declination measurements of Mars and fixed stars in Cayenne, Paris, and other places, Cassini found only three sets which, when compared, produced results leading to a horizontal parallax of Mars of 25″ by a dubious averaging technique. The answer in the case of Mars's parallax was certainly not unambiguous. Settling on the value of 25⅓″ obtained by the declination method, Cassini calculated the horizontal solar parallax to be 9½″, corresponding to a mean solar distance of 21,600 e.r.[58] He commented as follows on the accuracy of these figures:[59]

> But it is almost impossible to be certain of 2 or 3 seconds in the total parallax of Mars drawn from the comparison of several observations of which each one is subject to some imperceptible error. Moreover, a variation of 3 seconds in the total parallax of Mars is enough to cause a variation of 1,000 semidiameters of the Earth in his distance, even when he is closest to the Earth. From this it appears that it is no small undertaking to determine his least distance from the Earth to the nearest 1,000 semidiameters of the Earth and, consequently, that of the Sun to the nearest 2,000 or 3,000 semidiameters.

Cassini wrote this in 1684, when finally, after more than a decade's delay, he published a report on the expedition to Cayenne. Perhaps the reason for this long delay was the difficulty of supporting his initial announcement in 1673 about the parallaxes of Mars and the Sun. Historians have tended to accept Cassini's measurements and his analysis of them, crediting him and often also Flamsteed, with the first determination of a realistic solar parallax, even though some admit that the error margin was high.[60] From the foregoing it is clear, however, that the body of measurements of Mars made in connection with the expedition to Cayenne was worthless for detection of an actual parallax. The largest planetary parallax was still swamped by the errors of the measuring arcs.

The accuracy of Flamsteed's micrometrical measurements is more difficult to assess. The parallactic difference was of the order of 10″, twice the error margin that he himself assigned to his micrometer. His calculations added other errors, and his resulting parallax of Mars, calculated from the same few measurements, varied from 30″ to 15″. It is just possible that Mars's parallax was barely coming into the range of accuracy of the micrometer, but that the measurements were compromised by other uncertainties.

Yet, the measurements of Mars did show that the traditional notions of planetary distances were completely inadequate. If a parallax of less than 30″ was hidden in the error margin of the instruments, a parallax of, say 5′, which followed from the horizontal solar parallax of 2′21″ taken from Boulliau by Vincent Wing (1619–68) in his *Astronomia Britannica* of 1669,[61] would surely have been detected by the measurements of Mars during the opposition of 1672. Wing was, however, hopelessly out of touch on this issue; Kepler had already found that Tycho's observations did not support a solar parallax larger than about 1′ (p. 89).

The expedition of Richer to Cayenne was a great success, for it accomplished its main goal. Besides the unexpected and important dividend of the discovery of the varying length of a seconds pendulum,[62] Richer produced meridian measurements of the Sun that proved that a combination of vanishing small solar parallaxes and refractions continuing right up to the zenith gave consistent figures for the obliquity of the ecliptic at different latitudes. This result put a ceiling of perhaps 12″ of 15″ on the horizontal solar parallax, and this ceiling was supported by the attempted measurements of the parallaxes of Mars, which were still masked by the decreasing errors of the new generation of measuring instruments.

* 13 *
The New Consensus and Halley's Legacy

Although publication by Flamsteed and Cassini in 1673 of a solar parallax of 9.5″ or 10″ was a watershed, it did not end measurement of, and speculation about, the length of the astronomical unit. Neither did it lead immediately to a consensus in the scientific community. Over the next half-century, the most severe critic of their method, Edmond Halley (1656–1742), slowly replaced Flamsteed and Cassini as the most authoritative voice on the measurement of the astronomical unit.

Halley's first major astronomical project, for which he interrupted his studies at Oxford, was an expedition to the island of St. Helena in the South Atlantic from 1676 to 1678 for the purpose of mapping the southern stars. On St. Helena, Halley observed the transit of Mercury of 7 November 1677.[1] Up to this point, the main attraction of transits had been the accurate determination of apparent diameters and the improvement of the planetary theories. In 1663, however, the mathematician James Gregory (1638–75) had pointed out how useful transit observations could be for determining parallaxes.[2] Halley was therefore aware of this possibility. He was also lucky enough to be the first astronomer to observe both Mercury's initial entry on the Sun's disk and its final exit.[3]

On his return to England in 1678, Halley found that the astronomer Jean Charles Gallet in Avignon had been the only one in Europe to witness the transit.[4] From a comparison of Gallet's observation with his own, Halley calculated that Mercury's parallax was 21″ greater than the Sun's. From the known relative distances of the two bodies, it followed that Mercury's horizontal parallax was 1′6″ and the Sun's 45″. Halley, however, did not think that these values were very accurate.[5]

In his report on the transit, published as an appendix to his *Catalogus Stellarum Australium* in 1678, Halley discussed the various methods of finding solar parallax. He pointed out that in measuring Mercury's path across the Sun, all measurements, including that of the Sun's apparent diameter, could be made with a micrometer. All the observer had to do was to establish Mercury's path across the Sun and find the planet's smallest distance from the Sun's center. Comparison of the results from observers separated by a suitably great distance would yield a fairly accurate measure of the planet's parallax. Mercury's parallax was, however, not much larger than the Sun's, and this affected the accuracy of finding the latter from the former.[6]

Although Mars's parallax at opposition is twice as large as the Sun's, Halley felt that this method was liable to lead to errors. Apparently still unaware of the method used by Cassini and his French colleagues, Halley addressed himself only to the method used by Flamsteed in 1672. The measurements of Mars's distances from fixed stars required great care, a long telescope, and a precisely made micrometer. Even then, one could hardly get to the bottom of the matter. Therefore, in his opinion:[7]

> There remains but one observation by which one can resolve the problem of the Sun's distance from the Earth, and that advantage is reserved for the astronomers of the following century, to wit, when Venus will pass across the disk of the Sun, which will occur only in the year 1761 on 26 May, o.s. For if the parallax of Venus on the Sun is then observed by the method I have just explained, it will be almost three times greater than the Sun's, and the observations required for this are the easiest of all, so that through this phenomenon men can instruct themselves of all they could wish on that occasion.

Halley had little use for the methods that had been used up to that point, including the various methods of measuring Mars's parallax. Astronomers would simply have to wait nearly a century before the issue could be settled definitively. For an astronomer who had just returned from making exact position measurements of heavenly bodies, Halley showed remarkably little sensitivity to the related problems of refraction and solar parallax or even to the accuracy of the new generation of instruments that he himself had used.

In defending his own rather large solar parallax found from the transit of Mercury, Halley pointed out that those astronomers who made the Sun's parallax much smaller had only "reasons of probability" to support their opinions. As the principal of these reasons, he referred to Thomas Streete's harmonic argument of 1661 and gave it his own twist: if solar parallax were as small as 10″, Venus, a planet without a moon, would be larger than the Earth, a planet with a moon; if it were as large as 20″, Mercury, a primary planet, would be smaller than the Moon, a secondary planet. Since both cases would be unseemly, solar parallax had to be between 10″ and 20″.[8] This was, of course, not at all the principal reason why Cassini accepted a very small solar parallax.

Halley had frequently assisted Flamsteed in observations and he must have been fully aware that the new measuring arcs with telescopic sights were more accurate than Tycho's instruments. Yet, he cited Tycho's, not Cassini's or Flamsteed's, measurement of Mars's diurnal parallax as the one that set the upper limit on the Sun's parallax. Since Tycho had not been able to discern Mars's parallax, it had to be 1′ or less, and therefore the Sun's parallax had to be 25″ or less, "which I believe approaches the truth closely, after having with mature consideration weighed all its circumstances."[9] What were those circumstances?[10]

> I know from my own experience that [this solar parallax of 25″] can be neglected without scruple in the eclipses of the Sun and Moon, as such a small angle is not distinguishable in most astronomical instruments.

The question, however, was not whether an angle of 25″ was indistinguishable in *most* instruments; the question was whether it was distinguishable in the *best* instruments, those of the observatories at Paris and Greenwich. If Mars's parallax was indetectable with instruments at the Royal Observatory in Paris that erred at most, say 20″, this limited the solar parallax to 10″, regardless of the error margin Halley ascribed to "most astronomical instruments." In science, standards of precision are not set by average practice but rather by best practice.

Shortly after he printed this, Halley actually helped Flamsteed make a parallax measurement. Since he had moved into the new observatory at Greenwich in 1675, Flamsteed had on several occasions tried again to measure the parallax of Mars during an opposition. Presumably because no stars would be close enough to Mars for micrometer measurements on these occasions, he used "a sextant furnished with telescopicall sight & a very convenient movement" for these observations.[11] The opposition of January 1679 presented a very good opportunity. On 24 January, Mars was only 6° from opposition and its altitude was high enough to lift it above "ye reach of unlimited & doubtfull refractions."[12] Halley helped Flamsteed measure the evening and morning distances of Mars from several stars in Leo and Ursus Major. Flamsteed reported that Halley "alwayes concluded ye measures accurate."[13] From its distances from two stars, Flamsteed found Mars's diurnal parallax to be 15″, while by its distances from a third, he found it to be negative. Since this last result was obviously an error, he ignored it; the first two gave a horizontal solar parallax of 10″ or 12″.[14]

In his second Gresham Lecture of May 1681, Flamsteed drew the following conclusion from all his measurements of Mars's parallax:[15]

> . . . I conceave or observations confirme themselves sufficiently by so neare a consent[.] certeinly hence wee may conclude that the suns Horizontall parallax is not more yn ¼ of a minute[.] lesse it may . . . justly be supposed[;] nay I dare affirme with Wendelin[,] who with or Mr Horrox & the Author of ye Caroline Tables [Thomas Streete] supposes it no more then 15″ seconds[,] that the hypothesis of no solar parallax cannot be convicted of a sensible error.

And thus, although in print Flamsteed was on record as having determined the Sun's parallax to be a definite value, 10″, in this lecture he drew a more proper conclusion from all his measurements since 1672. The horizontal parallax of the Sun was less than the accuracy of the instruments, and if one assumed it to be zero in astronomical calculations, no measurable errors would be introduced. In spite of assisting in this measurement, or perhaps because of it, Halley never closed ranks with Flamsteed and Cassini on an official solar parallax of 9.5″ or 10″.

Perhaps it was Halley's influence that helped Newton ignore Flamsteed's solar parallax for several decades. Newton had, in fact, already developed his own ideas about solar parallax before the measurements of 1672. Some time between 1667 and 1671,[16] in testing the inverse square law of universal gravitation on the Moon's motion, he calculated the tendencies of the Moon to recede from the Earth and of the Earth to recede from the Sun. He argued that:[17]

> . . . if [the Moon's] tendency to recede acts so that it always faces the Earth with the same aspect, the tendency of this system of Earth and Moon to recede from the Sun ought to be less than the tendency of the Moon to recede from the Earth: otherwise the Moon would face toward the Sun, rather than toward the Earth.

Newton had already found the formula for this centrifugal tendency: the radius of motion multiplied by the square of the angular speed. Now, in the time the Earth completes one revolution around the Sun, the Moon completes 13.369 revolutions around the Earth. Setting the Earth-Sun distance equal to 100,000, and calling the Earth-Moon distance y, the tendencies to recede are to each other as $100{,}000 \times (1)^2$ is to $y \times (13.369)^2$. Therefore, $178.73y = 100{,}000$. If the tendencies were equal, y would be exactly 559.5, but since the Moon always faces the Earth, not the Sun, $y > 559.5$. This meant that in units in which the Sun's distance was 100,000, the Moon's distance had to be at least 559.5. Knowing the actual distance of the Moon, 60 e.r., Newton concluded that the Earth's radius had to be greater than 559.5/60 units, or 9.325 units. The solar parallax was, therefore, at least $\sin^{-1} 9.325/100{,}000$, or 19″.[18] In a letter written in 1673, he made the Sun's parallax at least 20″.[19] And although 20″ was a *lower* limit in these calculations, it remained Newton's preferred value for two decades, even when it no longer represented a lower limit in his mind.

Just how firmly this figure was fixed in Newton's mind is shown in his *Principia*, published fifteen years after Flamsteed's measurement. In book 3, "The System of the World," Newton calculated, where possible, the weights that bodies of equal mass would have at equal distances from the Sun and the planets, what their weights would be at the surfaces of these bodies, and what the densities of these bodies were. These calculations were possible only for those bodies that were centers of motion of other bodies, that is the Sun, Earth, Jupiter, and Saturn. For the first calculation Newton wished to compare the distances of (1) Venus from the Sun, (2) the Moon from the Earth, (3) the outermost Galilean moon from Jupiter, and (4) the Huygenian moon (Titan) from Saturn. To compare these distances, however, he needed a solar parallax, and for this he used his old value of 20″.[20] The result was that the weights of bodies of equal mass at equal distances from the centers of the Sun, Jupiter, Saturn, and the Earth were to each other as 1, 1/1,100, 1/2,360, and 1/28,700.[21]

In order to calculate the weights of these bodies of equal mass at the surfaces of the Sun, Jupiter, Saturn, and the Earth, Newton needed to know the radii of these bodies in terms of the Earth's radius, and thus he needed their apparent diameters. Flamsteed obligingly sent Newton a copy of the little manuscript tract on the diameters of the planets.[22] According to Flamsteed's figures, Jupiter's and Saturn's radii seen from the Sun were 19¾″ and 11″. Newton corrected this last figure to 10″ or 9″ on the specious grounds that "the globe of Saturn is somewhat dilated by the unequal refrangibility of light."[23] His real reason will become apparent presently. The Earth's apparent radius, seen from the Sun, was 20″, Newton's solar parallax. The radii of the Sun, Jupiter, Saturn, and the Earth, then, were to each other as 10,000, 1,063, 889, and 208, respectively. Knowing what the weights of bodies of equal mass were at equal distances from their centers, Newton used the inverse square law to calculate what bodies of equal mass would weigh on the surfaces of the Sun, Jupiter, Saturn, and the Earth. The weights would be in the ratio 10,000, 804½, 536, and 805½.[24]

What about Mercury, Venus, and Mars, however? Newton could not make such calculations for them, because no moons had been discovered about them. Was there perhaps a pattern that could be extended to include these planets? If one compared the relative weights of bodies of equal mass at the surfaces of Jupiter, Saturn, and the Earth—804½, 536, and 805½—with the apparent radii of these bodies as seen from the Sun—19¾″, 10″ or 9″, and 20″—the weights were very nearly in the same proportions as the square roots of the apparent radii (4.44, 3.00, 4.47), provided of course that Saturn's apparent radius seen from the Sun was 9″, not 11″ as Flamsteed had measured them.[25] Newton wrote:[26]

> Therefore the weights of equal bodies at the surfaces of the Earth and the planets are in about the half ratio of the apparent diameters seen from the Sun. There is as yet no certainty about the Earth's diameter seen from the Sun. I have taken it to be 40″, because the observations of Kepler, Riccioli, and Wendelin do not allow it to be much larger; the observations of Horrocks and Flamsteed seem to make it somewhat smaller. And I have chosen to err on the large side because if perhaps that diameter and [consequently] the gravity at the Earth's surface are the means among the diameters of the planets and the gravities at their surfaces, then since the diameters of Saturn, Jupiter, Mars, Venus, and Mercury, seen from the Sun, are about 24″

> and thus the parallax of the Sun about 12″, nearly as Horrocks and Flamsteed have determined it. But a slightly larger diameter agrees better with the rule of this corollary.

Although he felt free to take only slight liberties with the apparent diameters which Flamsteed supplied them, Newton almost ignored Flamsteed's value for the solar parallax. He trusted his own figure derived from the dynamics of the Sun-Earth-Moon system more than Flamsteed's figure, never even mentioning Cassini, and he used Keplerian harmonies that were supported by his parallax. Clearly, there was as yet no sign of a consensus on solar parallax in England.

In France the situation was no different. Cassini and his colleagues, of course, continued their attempts to measure parallaxes by the diurnal method. He and Picard used the method of measuring diurnal differences in right ascension by means of a pendulum clock on the comet of 1680 as well as on Venus, when its declination equaled the Sun's in the spring of 1681.[27] In the latter case Cassini found a solar distance of 22,000 e.r., as he had in 1672, and he claimed that Picard's independent measurements supported his own conclusion entirely.[28]

Cassini more than anyone else before 1700 took advantage of the new mode of disseminating scientific knowledge, the "paper" in a scientific periodical. In this manner he brought his new discoveries to the attention of the learned world very quickly, while at the same time establishing his priority. Yet, it was more than a decade after Richer's return from Cayenne that Cassini published his evaluation of the astronomical treasure. The reason for the delay, we may surmise, was Cassini's quandary about the results of the parallax measurements.

In his *Élémens de l'Astronomie verifiés par le Rapport des Tables aux Observations de M. Richer, faites en l'isle de Caïenne*, published in 1684, he could point with pride to the incontestable proof that his hypothesis of refractions and parallaxes had been proved correct beyond a shadow of a doubt by Richer's observations. Solar parallax, therefore, had to be very small. But in showing just how small it was, or rather, how small he thought it ought to be, Cassini did violence to the data, and he admitted as much on several occasions. In his tract on the comet of 1680, published in 1681, he noted that parallax observations "made by different astronomers, with different instruments, and by different methods, can have differences among them which could be taken for parallax."[29] In a lecture to the Academy in 1684, he admitted that comparisons of the observations of Mars in Cayenne and France only showed that Mars's parallax was insensible.[30]

Yet he did not abandon his efforts to tease a value for the solar parallax out of the observations made in 1672. As we have seen (pp. 140–42), in his 1684 report he coaxed the desired result from three sets of carefully selected meridian heights of Mars and a fixed star taken at Cayenne and Paris. In a further report in 1693 he even turned a null result in another set of the same measurements to his advantage. He argued that the lack of a difference was due to instrumental errors. Since no sensible error could have occurred in the Paris measurements, and since the Cayenne measurements could not have erred by more than 15″, this error margin contained Mars's parallax due to the latitude difference between Cayenne and Paris. Mars's horizontal parallax was, therefore, at most 25″, and the Sun's horizontal parallax was no more than 9″.[31]

Until he was rivaled and perhaps surpassed by Flamsteed in the 1680s, Cassini was the dominant figure in the new precision astronomy. His solar parallax of 9.5″ therefore carried great weight. In his own observatory, however, there was considerable disagreement. Picard, who was a part of the French team that observed Mars in 1672, concluded from a comparison of his own observations with those of Richer in Cayenne that the parallax of Mars, and therefore the parallax of the Sun, was insensible. Had there been a sensible parallax, it would surely have been detected on this occasion.[32] Roemer, also a part of the team in 1672, announced his determination of the speed of light in 1676, and in his demonstration used a solar parallax of at most 15″.[33] And Philippe de La Hire (1640–1718), who moved into the Royal Observatory a year after Roemer's return to Denmark in 1681, had also made observations of Mars in 1672. From many evening and morning distances measured from a number of stars in Aquarius between 22 September and 29 October, he had obtained results so inconsistent that he concluded that Mars's parallax was hidden by instrumental uncertainties. The parallax of the Sun could, therefore, be neglected. Thereafter, La Hire used a horizontal solar parallax of 6″ in his calculations; he did not change his mind after he obtained access to the better instruments of the Royal Observatory.[34]

In Rome, the Abbot and astronomer Francesco Bianchini (1662–1729) read Cassini's *Élémens de l'Astronomie* and decided to make his own measurements of Mars's parallax during the planet's opposition in 1685. He measured the time lapse in right ascension between Mars and a fixed star as they traversed the field of his telescope. His measurements, made from 30 May to 1 June, the days immediately following the opposition, produced a horizontal parallax of Mars of 39″.[35] Bianchini concluded:[36]

> If you assume a parallax of 40″, in round numbers, his distance from the Earth will be more than 5,100 [terrestrial] semidiameters. As you see, I have not troubled myself about the smallest fractions in order not to be considered uselessly fastidious in so uncertain a matter.

The problem of solar parallax, and therefore the problem of how large and how far away the heavenly bodies were, had thus hardly been solved in a definitive way by 1685. Among astronomers of the first rank, there was a rough consensus that the horizontal solar parallax was at most 15″, but there was virtually no agreement on a specific value. The measurements of Mars's parallax had been disappointing, and Huygens used a nice touch of understatement when he wrote to a provincial professor "that these conclusions for the distance of Mars are not nearly as certain or definite as those which put the Moon at [a distance of] 30 diameters of the Earth."[37]

If the community of practicing astronomers projected no consensus, and therefore no orthodoxy, to the learned or even merely educated community, what could one expect from popularizers? In his *Entretiens sur la Pluralité des Mondes* of 1686, Fontenelle, eager to impress his readers with the breathtaking vistas of the Copernican universe, still used Ptolemy's sizes and distances![38] Fontenelle went on to become one of the most important spokesmen for science, and his *Entretiens* went through numerous editions and printings. Yet in these numbers he made no changes until the sixth edition, printed in 1708, when he had been Perpetual Secretary of the Académie Royale des Sciences for over a decade.[39]

In the 1690s, however, a consensus did begin to emerge. It started with Huygens's comments on Newton's *Principia*. Huygens still clung to the apparent diameters and solar distances that he had published three decades earlier in *Systema Saturnium*. When he read the *Principia*, he recalculated the gravity at Jupiter's surface, according to Newton's method but with his own larger apparent diameters, and, surprisingly, Cassini's solar parallax of 10″ instead of his own of 8.2″.[40] He concluded that the weights of objects of equal mass at the surfaces of Jupiter and the Earth were to each other as 13:10, not 804 ½:805 ½ as Newton had calculated.[41] When he published his *Traité de la Lumière* with his *Discours de la Cause de la Pesanteur* in 1690, he took the opportunity to add some thoughts on Newton's *Principia* at the end. He wrote here about Newton's calculations for the gravity on other planets:[42]

> It is true that there is some uncertainty because of the distance of the Sun, which is not sufficiently well known, and which has been taken [by Newton] in this calculation as about 5,000 diameters of the Earth, while according to the dimension of Mr. Cassini it is about 10,000, which closely approaches what I have found formerly by probable reasons in my System of Saturn, to wit, 12,000.

Huygens's point was not lost on Newton, who called his critique of the *Principia* brilliant, and added: "The Parallax of the Sun is less than I concluded it to be."[43] Shortly afterwards, in 1692, when the clergyman and philosopher Richard Bentley (1662–1742) used a solar distance of 7,000 e.r. in his sermons against atheism,[44] Newton advised him (expressing all measures in terms of diameters):[45]

> Your assuming ye *Orbis Magnus* 7,000 diameters of ye earth wide [i.e., in diameter] implies ye Sun's horizontal Parallax to

> be half a minute. Flamsteed & Cassini have of late observed it to be but about 10″, & thus ye *Orbis magnus* must be 21,000 or in a rounder number 20,000 diameters of ye earth wide. Either assumption will do well & I think it not worth your while to alter your numbers.

Although for Bentley's particular purposes either "assumption" was acceptable, Newton implied here that a solar distance of about 20,000 e.r., corresponding to a solar parallax of about 10″, was now the preferred figure among scientists. Bentley, anxious to be scientifically "sound" in his arguments, made the appropriate changes in the published version of his sermons.[46]

By 1692 or 1693 Newton had thus abandoned his own figure for the solar parallax in favor of the value of Flamsteed and Cassini. When he allowed David Gregory (1661–1708) to publish his lunar theory in 1702, in which he "set down 10″ for the Sun's Horizontal Parallax,"[47] he publicly closed ranks with Flamsteed and Cassini.

Newton's successor in the Lucasian chair at Cambridge, William Whiston (1667–1752), had paraphrased Newton's ideas in his *New Theory of the Earth* of 1696. Whiston put the Sun's parallax somewhere between 12″ and 20″, and chose a mean figure of 15″.[48] In his astronomical lectures delivered at Cambridge in 1701, he reviewed all methods of determining the solar parallax, showing that those of Aristarchus and Ptolemy were useless. While Whiston did not flatly reject Horrocks's not altogether improbable conjecture, taken from "a certain Harmonical Magnitude of the Planets," the best method, he thought, was that of Mars's parallax at opposition. He reported and used Flamsteed's and Cassini's result of 10″ for the Sun's horizontal parallax.[49]

If Newton now lent his authority to the solar parallax of Flamsteed and Cassini, Halley did not. Halley's keen interest in parallax was now two decades old. He had observed the transit of Mercury of 1677 and had tried to determine Mercury's parallax from these observations. In 1679 he had helped Flamsteed measure Mars's diurnal parallax, and his trip to visit Hevelius in Gdansk had taught him important lessons about instrumental accuracy and measurement techniques. Halley was, therefore, in a good position to judge the accuracy of the parallaxes measured by his colleagues. Neither method, he thought, was accurate enough. When, in 1691, he presented his calculations of past and future times of transit of Mercury and Venus, he again indicated that Venus observed from different vantage points during a transit offered the only hope.[50]

Halley's influence can be seen in the opinions of David Gregory who, no doubt partially out of loyalty to his uncle James Gregory who had called attention to transits for this purpose in 1663 (see p. 144), proclaimed transits of Venus the only suitable occasions for measuring parallax. In his *Astronomiae Physicae & Geometricae Elementa* of 1702, Gregory wrote that for an accurate solar parallax the learned world would have to wait until the transit of Venus predicted for 1761. In the meantime, it seemed best to him to accept Newton's speculation of 1687, that is, that the Earth's apparent diameter as seen from the Sun was a mean or average between those of the other planets. This would produce a solar parallax of 12″.[51] Yet, in the same work Gregory printed Newton's lunar theory in which Newton took 10″ as his parallax.[52]

On the Continent, Huygens remained an important voice, and although in his critique of the *Principia* he had used Cassini's solar parallax, he never rejected his own of 8.2″ first published in 1659. In his *Cosmotheoros*, written during the last few years of his life and published posthumously in 1698, he used his own solar distance of 24,000 e.r. in all his calculations. He faithfully reported, however, the figure determined by Cassini and Flamsteed, based on measurements however inaccurate, beside his own figure, based on speculations however probable.[53] *Cosmotheoros*, quickly translated into English, French, and Dutch, brought the measurements of the astronomers of the French and English kings to the attention of a wide audience.

Cassini himself passed his ideas on to his son Jacques (1677–1756) and his son-in-law Giacomo Felippo Maraldi (1665–1729), who both followed in his astronomical footsteps. In 1704 Maraldi took advantage of another favorable opposition of Mars to determine its parallax. Using the now standard method of timing the intervals in right ascension between Mars and one or more fixed stars, Maraldi found time discrepancies of the order of 1 sec. (=15″ of arc) attributable to Mars's diurnal parallax in right ascension. From these he managed to find Mars's horizontal parallax to be about 24″ and the Sun's 10″, agreeing almost exactly with the results of the elder Cassini found three decades earlier.[54] It was, no doubt, on the basis of this agreement that Fontenelle used Cassini's and Flamsteed's and now Maraldi's solar distance in the sixth edition of his *Entretiens*, published in 1708. Having arrived at the Sun, he now told his pupil: ". . . we have made a journey of thirty-three million leagues."[55]

Fontenelle can be forgiven for not putting qualifiers in this and other like statements. They would have diminished the shock value which was one of the main attractions of his popular tract. It will not do for us, on the other hand, to conclude that Cassini had measured the distance of Mars "with an accuracy never before attained" or that "he was the first to remove the doubts and inaccuracies which had clung for so many centuries to these fundamental data."[56] There is an important difference between saying that Cassini's solar distance was within about 1″ from the modern value and that it was *accurate* to within that amount. If there were no difference, what would be wrong with saying that Huygens *guessed* the solar parallax with an accuracy of ½″?

Although Cassini and his contemporaries had great difficulties in estimating the accuracies of their instruments and measurements and had no error theory to tell them how meaningful and how accurate the results of their calculations were, Cassini took some pains to not knowingly overstate what he considered to be the accuracy of his conclusions. A solar parallax of 9½″ would mean a solar distance of 21,600 e.r. The accuracy of the latter figure depended on the accuracy of the former, but since, as Cassini felt, it was almost impossible to be certain of 2″ or 3″ in the parallax of Mars, it was no mean feat to determine the Sun's distance to 2,000 or 3,000 e.r. The difficulty of determining such large distances with the same accuracy increases with the square of the distances, so that a distance 20 times greater is 400 times more difficult to determine with the same accuracy. Cassini continued:[57]

> This remark is so much the more necessary because some suppose that the distances of the stars can be measured with the same ease and the same accuracy as those with which we measure the distances of inaccessible places on the surface of the Earth. And they set forth the distances of the farthest planets and even those of the fixed stars in leagues and miles, as we set forth the distances between cities. It would be no mean feat to know them to within some millions of miles. Thus, since the distance of the Sun from the Earth approaches 22,000 semidiameters of the Earth, and since the Earth's semidiameter is commonly given as 1,500 leagues, one can say that the distance of the Sun from the Earth is about 33 million leagues without being answerable for a discrepancy of one or two million, just as on Earth one would not be answerable for one or two leagues in a distance of 32 or 33 leagues when it is judged only by estimate.

If we can agree, then, that by 1700 a horizontal solar parallax of about 10″ and a solar distance of about 20,000 e.r. were the preferred figures, we should not let the close agreement of these values with the modern ones seduce us into drawing unwarranted conclusions about their accuracy. If the solar parallax of Cassini and Flamsteed had, in fact, been as *accurate* as has sometimes been claimed by historians, we would expect that their successors, with even more accurate instruments and better techniques, would measure it even more accurately and closer to the modern value. From 1672 until the Venus transit measurements of 1761, the astronomical unit was usually found by measuring Mars's parallax at opposition. We would expect, therefore, that during this period the consensus would become ever wider while the consensus value for solar parallax would approach ever closer to the modern value. In fact, the opposite happened.

Newton changed his mind on solar parallax several times, albeit within rather close limits. While in the second edition of the *Principia*, printed in 1713, he used Cassini's and Flamsteed's 10″,[58] in the notes for the third edition the figures 11″, 12″, and 13″ are found.[59] In the second English edition of the *Opticks* (1717) he used 12″.[60] In 1719 James Pound (1669–1724) and his nephew James Bradley (1692–1762) measured Mars's parallax at opposition, and their measurements supported a solar parallax of not less than 9″ and not more than 12″.[61] In the third edition of his *Principia* (1726), Newton used the average between these two limits, 10½″.[62]

Halley had remained unimpressed with these measurements, and in 1716 he published his best known call to arms for the 1761 transit of Venus. Here he argued that the accuracy of this parallax measurement would be one part in 500. He did not even mention Flamsteed, Cassini, or the parallax of Mars.[63] Until 1761 astronomers were given no hope of determining the solar parallax accurately. The only interim solution offered was to pick the fairly unobjectionable value of 12½″, a figure that would insure that the Earth, a planet with a moon, was larger than Venus, a planet without a moon, and that Mercury, a primary planet, was larger than the Moon, a secondary planet.[64] Halley's influence on this issue grew as the transit of Venus approached.

Over the next four decades (1720–60), successive measurements of Mars's parallax at opposition, and even Mercury's parallax with respect to the Sun during a transit, made the Sun's horizontal parallax larger, not smaller. The researches of Jacques Cassini in 1736 and Nicolas Louis de Lacaille (1713–62) in 1751 resulted in values from 11″ to 15″, with very wide scatters.[65] On the eve of the expeditions to observe the transits of Venus, the astronomer Joseph Nicholas Delisle (1688–1768) reviewed all determinations of solar parallax since 1672 and concluded that the Sun's horizontal parallax was between 10″ and 14″, "so that the opinion of Mr. Halley, who makes it 12″, is not very far from the true parallax."[66] If the expeditions to observe the transits of Venus of 1761 and 1769 did not lead to a solar parallax accurate to one part in 500, as Halley had hoped, they did produce a range of values much narrower and vastly superior to anything produced up to that time.

Yet, even if we conclude that before 1761 astronomers were not so much measuring planetary parallaxes as the accuracies of their instruments and techniques, we must not lose sight of the fact that by the end of the seventeenth century a consensus had been reached which made the Sun's horizontal parallax at most 15″ and its distance at least 55,000,000 miles. The solar system was thus enormous by the standards of the beginning of that century. Using a solar parallax of 10″ (corresponding to a solar distance of about 20,500 e.r.), Whiston gave the dimensions shown in table 19 in 1715.[67]

These sizes and distances were one of the marvels of the new astronomy. Not a few people alive in 1700 had first learned their cosmic dimensions from the Ptolemaic scheme, in which the Sun was 5½ times as big in diameter as the Earth, and 1,200 e.r., or about 5,000,000 miles removed from it—20,000 e.r. had been the traditional radius of the entire cosmos![68] The sphere of Saturn was now about 200,000 e.r. in radius, and the solar system (as it was called by the Copernicans) did not end there: the elliptical path devised by Halley for the comet of 1682 placed its aph-

TABLE 19 Whiston's Scheme of Sizes and Distances

Body	Diameter in (English) Miles	Heliocentric Distance in (English) Miles
Moon	2,175	—
Sun	763,460	—
Mercury	4,240	32,000,000
Venus	7,906	59,000,000
Earth	7,935	81,000,000
Mars	4,444	123,000,000
Jupiter	81,155	424,000,000
Saturn	67,870	777,000,000

elion 4 times farther from the Sun than Saturn's aphelion. Comets with parabolic paths linked our region of space with other regions.

New illustrations, verbal as well as pictorial, were needed to explain, and at the same time stress, these amazing new magnitudes. In his *Cosmotheoros*, Huygens showed the relative sizes of the Sun and planets very effectively (see fig. 13), but it was more difficult to show these sizes in proportion to the planetary distances:[69]

> . . . it will be worth while to set before you at once [i.e., in one picture] . . . the Magnificence and Fabrick of the Solar System. Which we can't possibly do in so small a space as one of our Leaves will but admit of, because the Bodies of the Planets are so prodigiously small in comparison of their Orbs.

Huygens, therefore, had to make up a picture with words. Referring his readers to figure 13, he asked them to imagine a series of concentric circles,[70]

> whose outermost Circle representing the Orb of *Saturn*, must be conceived three hundred and sixty foot in Semidiameter. In which you must place the Globe and Ring of Saturn of that bigness as the . . . Figure shows you. Let all the other Planets be supposed every one in his own Orbit, and in the middle of all the Sun, of the same bigness that that Figure represents, namely, about four inches in Diameter. And then the Orbit or Circle in which the Earth moves, which the astronomers call the *magnus Orbis*, must have about six and thirty foot in Semidiameter. In which the Earth must be conceived moving, not bigger than a grain of Millet, and her Companion the Moon scarcely perceivable, moving round her in a Circle a little more than two Inches broad.

A bullet shot out of a gun with the enormous speed of a hundred fathoms per pulse beat, that is, about 600 ft/sec., would take 25 years to reach the Sun and 250 years to reach Saturn![71]

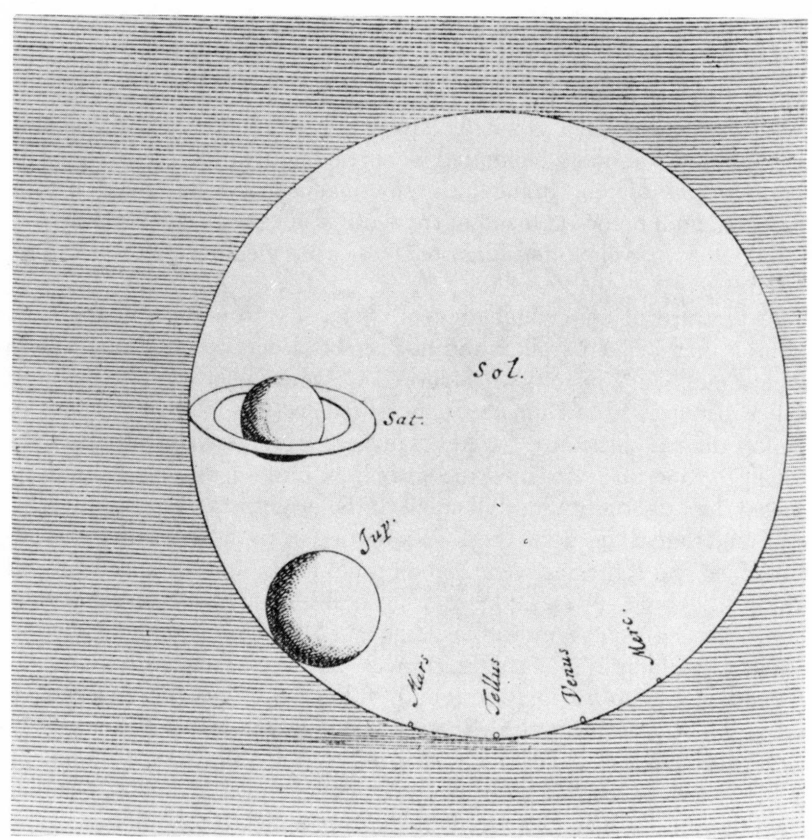

FIG. 13. Huygens's representation of the relative sizes of bodies in the solar system. From *Oeuvres complètes de Christiaan Huygens,* vol. 21, 1944; reprinted by permission of Hollandsche Maatschappij der Wetenschappen, Haarlem, the Netherlands.

And there still remained the fixed stars, thought by now by most astronomers to be suns "scatter'd and dispersed all over the immense spaces of the Heaven," and to be about as far from each other as from our Sun, which is so far that they appear "just like little shining Points."[72] At a time when the baseline of stellar parallax measurements (at least for Copernicans) had been expanded to about 40,000 e.r., or 160,000,000 miles, and telescopic sights had made instrumental accuracy much greater, there was justifiable hope that the annual parallax of the nearer, that is, brighter stars, would no longer escape the instruments. In 1669 Robert Hooke measured the zenith distances of γ Draconis in July, August, and October, and he found a difference of 24″ to 26″. He concluded that there was a sensible parallax, which constituted proof of the Earth's motion.[73] Hooke's measurements were greeted with skepticism. Picard measured the elevations of a bright star in Lyra at the solstices from 1675 to 1681 and reported no differences.[74] Flamsteed, however, after measuring the zenith distances of the polestar from 1689 to 1696, did find a seasonal variation of 42″ which he ascribed to parallax.[75] It turned out later that the variations found by Flamsteed were real enough, but were caused by the aberration of light. In the meantime, Jacques Cassini showed that the maxima and minima occurred at the wrong times for these variations to be caused by parallax, and Flamsteed's measurements were therefore rejected.[76]

At the turn of the eighteenth century most astronomers were of the opinion that stellar parallax had not yet been detected, although there were some, like Whiston, who ignored Cassini's objection and deduced stellar distances from Flamsteed's measurements.[77] One had to resort to conjecture and difficult estimates. Huygens progressively restricted the size of the aperture admitting the Sun's light into a lensless tube until he judged the brightness admitted equal to the brightness of Sirius at night. He found that at this point the 1/27, 664th part of the Sun's disk was admitted, so that if the Sun were 27,664 times its present distance removed from us it would appear equal in brightness and size to Sirius. If one assumed that Sirius was, indeed, as large and as bright as our Sun, then its distance would be 27,664 solar distances,[78] or, according to his own solar distance in *Cosmotheoros*, about 660,000,000 e.r. This guess was imaginative but very problematical, and it was not very influential.

Newton made a somewhat better conjecture, published posthumously in his *System of the World*. If we let Saturn's radius be r and its distance from the Sun R, then the area of the disk presented to the Sun is πr^2, and the total surface area of the "sphere" of Saturn is $4\pi R^2$. The proportion of the Sun's total light intercepted by Saturn is, therefore, $\frac{1}{4}(r/R)^2$. However, r/R is simply the sine of Saturn's angular radius seen from the Sun—9″ according to Newton (see p. 148). Hence, $r/R = 1/22{,}918$, and Saturn receives $1/(2.1 \times 10^9)$ of the Sun's total light. Now, of course, only half the Sun can shine on Saturn at one time, so that the planet receives $1/(1.05 \times 10^9)$ of the sunlight emanating from the hemisphere facing it. Moreover, Saturn does not reflect all the light it receives. Assume it only reflects $\frac{1}{4}$, then it reflects $1/(4.2 \times 10^9)$ of the sunlight emanating from one hemisphere. Since "light is rarified in the duplicate ratio of the distance," if the Sun were $10{,}000\sqrt{42}$, or about 65,000 times as far from Saturn as it now is, it would appear to us as bright as Saturn. Since Saturn is somewhat brighter than a fixed star of the first magnitude, Newton concluded that at a distance of about 100,000 times Saturn's distance the Sun would appear as bright as a first magnitude fixed star. Therefore a first magnitude fixed star equal to our Sun in size and intrinsic brightness had to be 100,000 times as far away from us as Saturn is, and it would have an annual parallax of 13‴.[79]

But Saturn's distance from the Earth was by Newton's own measure of the order of 200,000 e.r., so that the distance to the nearest fixed stars in terrestrial radii, let alone in miles, was awkward to express and almost impossible to comprehend. In the Middle Ages the then almost incomprehensible distance to the fixed stars, 20,000 e.r., had been illustrated by the calculation of how many years Adam would still have to walk, at a rate of 25 miles per day, to reach the fixed stars, had he started his journey on the day he was created (see pp. 37–38). Now a new illustration was needed. Huygens's cannon ball, traveling at 600 ft/sec. was adequate for distance within the solar system, but it would take 691,600 years to reach a fixed star which he had calculated to be 27,664 times as far away as the Sun.[80] Such a time span was, however, itself difficult to comprehend. The traditional Christian time frame, from creation to final judgment, was measured in a few thousands of years, not hundreds of thousands of years; the creation was still commonly put at about 4000 B.C. Newton's stellar distance would make the problem even worse. Not even the speed of a cannon ball would, therefore, suffice to express the distance of the stars in numbers the meaning of which could be grasped. In the long run, it was thus to be expected that the time taken by the fastest entity known, light, to cover these enormous distances should be used as an illustration and eventually as the measure itself. Already in 1694 in a paper in the *Philosophical Transactions*, a certain Francis Roberts calculated from a solar distance of 20,000 e.r. and a modest stellar distance of 6,000 times the solar distance,[81]

> That Light takes up more time in Travelling from the Stars to us, than we in making a *West India* Voyage (which is ordinarily performed in six Weeks).

* 14 *
Conclusion: Measurement, Theory, and Speculation

Besides enriching the Western tradition of learning with a complete system of mathematical astronomy, Ptolemy also gave it a detailed quantitative picture of the cosmos. He was quite right in separating his more rigorous mathematical planetary theories from his more speculative cosmic dimensions, putting the latter in a separate work, the *Planetary Hypotheses*. Unfortunately however, his discussion of the sizes of heavenly bodies and their spheres was somehow missing in the copies of the *Planetary Hypotheses* that survived in the original Greek; it also seems to have been missing at an early stage from many Arabic copies. Although Ptolemy's authorship of the scheme of sizes and distances was known by some Moslem writers, that knowledge was not passed on to the medieval Christian tradition. Not until the 1960s, therefore, was Ptolemy's intellectual property restored to him.

It is perhaps partly because it was a tradition of unknown origin that the scheme of cosmic dimensions that occupied an important place in Western thought from the thirteenth to the seventeenth century has received scant attention from modern historians. Had it been known that Ptolemy was its author, it would surely have been treated in connection with the Ptolemaic corpus of astronomy. As we have seen, Ptolemy's speculations about the sizes and distances of heavenly bodies took on a canonical status in the High Middle Ages, and his scheme of sizes and distances, or parts of it, can be found in medieval literature from the most popular to the most technical.

The Ptolemaic cosmic dimensions were not challenged until the sixteenth century, when the Copernican and Tychonic schemes of the cosmos dictated a new set of relative distances. Copernicus and Tycho did not, however, make significant changes in the solar distance, so the absolute distances in their systems were of the same order of magnitude as the Ptolemaic distances. While Copernicus had to make the sphere of the fixed stars very large indeed, the sphere of Saturn and everything in it actually shrunk. Only the telescope could change this.

The influence of the telescope on the sizes and distances in our solar system was direct in the case of the apparent diameters and indirect in the case of the distances. The planetary apparent diameters that had been unchallenged since Hipparchus and Ptolemy were discarded in the course of the seventeenth century. The telescope showed the planets as disks whose angular diameters could at first be estimated and later measured with the micrometer. The resulting apparent diameters were an order of magnitude smaller than the traditional ones. In the case of the Sun, the micrometer allowed measurements of the seasonal variations of its apparent diameter, and these measurements were sufficiently accurate to show that Kepler had been correct when he suggested that perhaps the solar eccentricity should be halved.

When it came to distances, the telescope's influence was less direct. The conviction that the parallax of Mars was actually measured in 1672 has been misplaced. As we have seen, an upper limit was placed on the Sun's parallax (meaning a lower limit on its distance) by a complicated interaction between solar theory (which could to some extent be verified by observation) and the combination of corrections for solar parallax and refraction. Cassini put this upper limit at about 10″; most other astronomers at the forefront of their profession at that time put the limit at something like 15″.

If Mars's parallax was still hidden by instrumental errors in 1672, then the question is: When *was* a parallax (other than the Moon's) first measured? Here we must go to the Venus transits of 1761 and 1769. Halley had thrown all his considerable prestige behind this project, and he was entirely correct. He argued on good grounds that simultaneous observations of a Venus transits would give a Venus parallax accurate to one part in 500. The famous expeditions did not, in fact, give a result with any such accuracy at all. But they did produce a solar parallax accurate to perhaps the nearest second of arc. This was a tremendous achievement that had lain totally outside the ken of the seventeenth century. The historical role of the Venus expeditions of 1761 and 1769 must, therefore, be strengthened. The result achieved by these expeditions was not eighteenth-century icing on a seventeenth-century cake, a more accurate measurement of the astronomical unit already roughly determined by Cassini and Flamsteed. It was the very first positive determination of a parallax other than the Moon's.

Yet, when all is said and done, we must also recognize that toward the end of the seventeenth century astronomers managed to put together a quantitative picture of the new universe that was changed very little by the Venus expeditions. The new distances came out of the interaction between theory and observation in two stages. The first was essentially a naked-eye stage. The *Rudolphine Tables* of Kepler were based on naked-eye measurements and elliptical astronomy. For those who cared to examine all the various dimensions of Kepler's astronomical work, it was clear by 1630 that a solar parallax of 3′ had become untenable and that an upper limit of 1′ was indicated. His arguments, however, had less impact than they might have had because Kepler made physical and harmonic arguments central in all his discussions of sizes and distances. Not many astronomers agreed with this approach.

The second stage in the emergence of a new consensus on distances was a result of the great increase in the accuracy of position measurements made possible by the telescopic sight and, to some extent, the micrometer.

If the expedition to Cayenne did not result in the positive measurement of Mars's parallax, it did show that the solar parallax was vanishingly small, that is, less than 15″. Such a small parallax and the new corrections for refractions it necessitated fit well with the demonstration, made possible by the micrometer, that the Sun's eccentricity should be halved.

This episode shows well the interaction between theory and instruments. One might have expected that a direct measurement of solar or planetary parallax with a new and more accurate instrument gave astronomy its new astronomical unit, while at the same time mending the residual errors in solar and planetary theory. What happened in fact was that a number of factors, instrumental and theoretical, interacted to produce a new upper limit on solar parallax. After choosing, perhaps subconsciously, a particular value near that limit, Giovanni Domenico Cassini produced that value from inconclusive measurements made during Mars's favorable opposition in 1672. The astronomical community thereupon slowly converged on this new solar parallax as a consensus, largely on the authority of Cassini and, to some degree, John Flamsteed. This consensus value, which gained strength and authority with every important convert, such as Newton, was confirmed periodically by measurements of Mars's parallax that were, in retrospect, as unreliable as the measurements of that planet made in 1672.

One astronomer, Edmond Halley, resisted the temptation to join this consensus. Halley's early notions concerning solar and planetary parallaxes were undistinguished. Assisting Flamsteed in his attempted measurements of Mars's diurnal parallax showed Halley, however, that the results of Flamsteed and Cassini were not much better than his own. For the rest of his life, he therefore rejected all measurements of Mars's parallax and the solar parallaxes derived from them. To his mind, the only hope of measuring not merely an upper limit, but an actual value for a planetary parallax was during the exceedingly rare transits of Venus. Having been born in 1656, he would surely not live to witness the next transit, scheduled for 1761.

Halley's prestige in the astronomical community grew only gradually, but after the deaths of Cassini in 1712 and Flamsteed in 1719, he was one of the profession's elder statesmen. Astronomer Royal since 1720, his repeated reminders that the upcoming transits presented rare and crucial opportunities for parallax measurements, and his proposal for how these measurements should be made, gained weight as the events approached. Halley died in 1742, but over the next two decades his fame increased spectacularly. In 1758 he became a popular hero when the great comet last seen in 1682 returned on schedule, as he had predicted. Three years later a large number of observers all over the globe observed the transit of Venus; in 1769 this international cooperative effort was repeated. The result of these expeditions was a positive determination of the solar parallax, as Halley had predicted.

One aspect of this story makes fertile ground for further investigation. Kepler injected an important stream of mathematical, or "harmonic," speculation into astronomy. The sizes of the planets, their heliocentric distances, yes even their densities, ought to stand in some simple mathematical relationship to each other. Although his immediate followers might not agree with him on the specific relationship, for example substituting diameters for volumes in the proportionality with distance, they agreed that these speculations had an important place in astronomy. But while for Kepler and, to a somewhat lesser degree, Horrocks such a priori proportions were part of a larger structure of mutually reinforcing components, by the second half of the century the proportions had been stripped of connections with other aspects of astronomy. Even though his own evidence denied it, Huygens felt that the further a planet is from the Sun the larger it ought to be, and he gave no further justification for this belief. Fifty years later, Halley could still argue, without any larger context, that Mercury, a primary planet, ought to be larger than the Moon, a secondary planet, and that the Earth, a planet with a moon, ought to be larger than Venus, a planet without a moon.

Although progressively limited by the findings of instruments, harmonic speculations thus continued to be used to fill gaps in our knowledge concerning the universe. If the a priori approach of Kepler lost its appeal in the eighteenth century, the a posteriori approach, also practiced by him, remained in use by astronomers. It produced Bode's "law" in the eighteenth century and Kirkwood's "gaps" in the nineteenth.

If the seventeenth century saw the birth of modern science, then surely the telescope is the prototype of modern scientific instruments. Without this instrument, which was the first extension of one of the human senses, there would have been no mountains on the Moon, no phases of Venus, no satellites around Jupiter and Saturn, no estimate of the speed of light, and no substantial improvement on the accuracy of Tycho Brahe's measuring instruments. Newton's sizes and distances of the Sun, Moon, and planets might very well have been the same as Copernicus's, at best the same as Kepler's.

The new quantitative picture of the solar system that came into focus toward the end of the seventeenth century was as fully integrated into theoretical astronomy as the traditional picture had been. The new sizes and distances represent the first major impact of precision telescopic astronomy on solar and planetary theory. They were also one of the wonders of the new age of science, an entirely new quantitative world picture to go along with an entirely new science.

Notes

ABBREVIATIONS

In citing works in the notes, short titles have generally been used; full citations are listed in the Bibliography. Works frequently cited have been identified by the following abbreviations.

Almagest: Ptolemy. *Syntaxis Mathematica*. Edited by J. L. Heiberg. Vol. 1 (in 2 parts, 1898 and 1893, respectively) *Claudii Ptolemaei Opera quae Exstant Omnia*, 3 vols. Leipzig: Teubner, 1893–1952.

Arabic Version: Bernard R. Goldstein. *The Arabic Version of Ptolemy's "Planetary Hypotheses." Transactions of the American Philosophical Society* 57, part 4 (1967).

Aristarchus: T. L. Heath. *Aristarchus of Samos the Ancient Copernicus*. Oxford: Clarendon Press, 1913.

Controversy: Stillman Drake and C. D. O'Malley, eds., trans. *The Controversy on the Comets of 1618*. Philadelphia, University of Pennsylvania Press, 1960.

De Rev.: Nicholas Copernicus, *De Revolutionibus Orbium Caelestium Libri Sex*. Vol. 2 *Nikolaus Kopernikus Gesamtausgabe*. 2 vols. Munich: Oldenbourg, 1944–49.

Dialogue: Galileo Galilei. *Dialogue concerning the Two Chief World Systems—Ptolemaic & Copernican*. Translated by Stillman Drake. Berkeley: University of California Press, 1967.

Discoveries: Stillman Drake, ed., trans. *Discoveries and Opinions of Galileo*. New York: Doubleday, 1957.

DSB: *Dictionary of Scientific Biography*. 16 vols. New York: Charles Scribner's Sons, 1970–80.

Duncan: Kepler, *Mysterium Cosmographicum: The Secret of the World*. Translated by Alistair M. Duncan. New York: Abaris Books, 1981.

Epitome: Johannes Kepler, *Epitome of Copernican Astronomy: IV and V*. Translated by Charles Glenn Wallis. In *Great Books of the Western World*, vol. 16, *Ptolemy, Copernicus, Kepler*, 845–1004. Chicago, London, and Toronto: Encyclopaedia Britannica, Inc., 1952.

HAMA: Otto Neugebauer. *A History of Ancient Mathematical Astronomy*. 3 parts. New York, Heidelberg, and Berlin: Springer-Verlag, 1975.

HOA: J. L. E. Dreyer. *History of the Planetary Systems from Thales to Kepler*. Cambridge: Cambridge University Press, 1906. Reprinted as *A History of Astronomy from Thales to Kepler*. New York: Dover, 1953.

JHOP: *Jeremiae Horrocci Opera Posthuma*. Edited by John Wallis. London, 1673.

JKAOO: *Joannis Kepleri Astronomi Opera Omnia*. Edited by C. Frisch. 8 vols. Frankfurt: Erlangen, 1858–70.

JKGW: *Johannes Kepler Gesammelte Werke*. Munich: C. H. Beck, 1937–.

Manitius: Ptolemy. *Handbuch der Astronomie*. Translated by K. Manitius. 2 vols. Leipzig: Teubner, 1912–13, 1963.

Mem.: Académie Royale des Sciences. *Mémoires de l'Académie Royale des Sciences depuis 1666 jusqu'à 1699*. 9 vols., numbered 3–11. Paris, 1729–32.

NCR: *Nicholas Copernicus on the Revolutions*. Translated by Edward Rosen. Baltimore: Johns Hopkins University Press, 1978.

OCCH: *Oeuvres Complètes de Christiaan Huygens*. 22 vols. The Hague: Martinus Nijhoff, 1888–1950.

OGG: *Le Opere di Galileo Galilei*. Edizione Nazionale. Edited by Antonio Favaro. Florence, 1890–1909; reprinted 1929–39, 1964–66.

Oldenburg: *The Correspondence of Henry Oldenburg*. Edited by A. R. Hall and M. B. Hall. Madison: University of Wisconsin Press and London: Mansell, 1963–.

PGOO: *Petri Gassendi Opera Omnia*. 6 vols. Lyons, 1658.

Phil. Trans.: *Philosophical Transactions*.

Registre: Académie Royale des Sciences. Archives. "Registre des procès-verbaux des séances."

TBOO: *Tychonis Brahe Opera Omnia*. Edited by J. L. E. Dreyer. 15 vols. Copenhagen, 1913–28; Reprint. Amsterdam: Swets and Zeitlinger, 1972.

Whatton: Horrocks, *The Transit of Venus across the Sun: a Translation of the Celebrated Discourse thereupon, by . . . Jeremiah Horrox . . . to which is prefixed a Memoir of his Life and Labours by. . . Arundell Blount Whatton*. London, 1859, 1868.

Chapter One

1. Compare, e.g., Nicolson's essays in *Science and Imagination* with the few pages Kuhn devotes to the role of the telescope in the Great Debate in *The Copernican Revolution*, 219–25.

2. Schickard, W. *Schickardi pars Responsi*, 9.

Chapter Two

1. Kahn, *Anaximander*, 55–68, 81–98. Kahn's interpretation of the sources is soundly refuted by Dicks in "Solstices, Equinoxes, and the Presocratics," 36. See also Dicks, *Early Greek Astronomy to Aristotle*, 44–46.

2. *HAMA*, 652–54.

3. Neugebauer states (*HAMA*, 653): "It is clear that neither the 'measurements' of distances nor the astronomical 'observations' are more than crude estimates, expressed in convenient round numbers." The first actual measurement of the Earth's circumference was made by the Moslem astronomer al-Bīrūnī in the tenth century A.D. See al-Bīrūnī, *The Determination of the Coordinates*, 177–80.

4. Diogenes, *Lives of the Eminent Philosophers*, ii, 8; Dicks, *Early Greek Astronomy to Aristotle*, 58.

5. *Aristarchus*, 351–411.

6. Ibid., 353.

7. Ibid. Note that Aristarchus did not use the sexagesimal system of dividing the circle. The division of the zodiac into 360° began around 150 B.C. among the Greeks and was used by Hipparchus. See *HAMA*, 590.

8. Neugebauer calls this value "a purely fictitious number" (*HAMA*, 642) and characterizes all Aristarchus's numerical data as "nothing but arithmetically convenient parameters, chosen without consideration for observational facts which would inevitably lead to unhandy numerical details" (*HAMA*, 643). Tannery argues that Aristarchus's method was, in fact, used first by Eudoxus, who may have assumed the angle subtended at the Sun by the Earth-Moon distance, $\angle ESM$, to be 6°, and that Aristarchus halved this value. See Tannery, *Recherches*, 225. For $\angle EMS = 6°$ one finds the ratio of distances and sizes 9½:1, which is not inconsistent with the 9:1 ratio of sizes ascribed to Eudoxus by Archimedes in his *Sand Reckoner* (*Works*, 223). But the supposition that Eudoxus used lunar dichotomy to arrive at this ratio is not supported by the evidence. See *HAMA*, 662.

9. *Aristarchus*, 377–81.

10. See also *HAMA*, 642–43.

11. *Aristarchus*, 383.

12. The apparent size of the Moon's disk is not always the same. Aristarchus here would rule out the possibility of an annular eclipse, in which the Moon's disk is centered on the Sun but not large enough to cover it entirely.

13. *Aristarchus*, 383–85.

14. Ibid., 353.

15. Ibid., 403–7.

16. Ibid., 407–9.

17. Ibid., 409–11.

18. Archimedes, *Works*, 223: ". . . Aristarchus discovered that the sun appeared to be about 1/720th part of the circle of the zodiac."

19. *Aristarchus*, 353: ". . . the moon subtends one-fifteenth part of a sign of the zodiac."

20. Ibid., 355–71.

21. See, e.g., R. R. Newton, *The Crime of Claudius Ptolemy*, 176–77, 392–94.

22. During the Hellenistic period the normal order of the planets in Greek horoscopes was Sun-Moon-Saturn-Jupiter-Mars-Venus-Mercury. The days of the week that we still use date from the same period and are derived from the order Moon-Mercury-Venus-Sun-Mars-Jupiter-Saturn. This order became the standard one after Ptolemy. See Neugebauer, *The Exact Sciences in Antiquity*, 169.

23. *HAMA*, 647–51; Osborne, "Archimedes on the Dimensions of the Cosmos."

24. Archimedes, *Works*, 226–32; *HAMA*, 643–47.

25. Heath gives only a brief paraphrase of Archimedes' text (*Works*, 223–24). Shapiro has incorporated an English translation of Archimedes' full text in "Archimedes's Measurement of the Sun's Apparent Diameter," 82–83.

26. In what follows I have relied on *HAMA*, 322–29; Swerdlow, "Hipparchus on the Distance of the Sun," and Toomer, "Hipparchus on the Distances." Toomer's "Hipparchus" in *DSB* is an excellent summary of current knowledge about Hipparchus.

27. Toomer, "Hipparchus," 211–13.

28. Toomer, "Hipparchus on the Distances," 133–38.

29. *HAMA*, 328–29; Toomer, "Hipparchus on the Distances," 131–38. To reconstruct Hipparchus's procedure and to arrive at his results, we must know the lunar declinations, which can only be found if the date of the eclipse is known.

Toomer took the six solar eclipses that were total in the Hellespont between 309 and 128 B.C., found the lunar declinations at the time of each eclipse, and used these to calculate the lunar distance. Only the eclipse of 189 B.C. gave Hipparchus's result.

30. Toomer, "Hipparchus," 213.

31. Although the general method was clear enough, Hipparchus's actual procedure had been the subject of confusion compounded by error. In "Hipparchus on the Distance of the Sun," Swerdlow has corrected the errors and has established that Hipparchus used the eclipse diagram to find the lunar distance after assuming a solar distance.

32. Toomer, "Hipparchus," 214–15; idem, "Hipparchus on the Distances," 129–31.

33. *Arabic Version*, book 1, part 2, 8.

34. *HAMA*, 654–55.

35. *HAMA*, 655–57.

Chapter Three

1. *HAMA*, 836–37.

2. The two lunar models are in books 4 and 5 of the *Almagest*. See also *HAMA*, 84–88; Pedersen, *Survey*, 184–87. The definitive work on the sizes and distances of the planets from Ptolemy through the Middle Ages is Swerdlow, *Ptolemy's Theory*. I am greatly indebted to Swerdlow's study for the materials in this and the following two chapters.

3. In sexagesimal numbers, the radius, R, of the deferent was 49;41 parts, the eccentricity, e, 10;19 parts, and the radius of the epicycle, r, 5;15 parts. According to the model, the maximum distance at syzygy is $R + e + r = 65;15$ parts; the minimum distance at syzygy is $R + e - r = 54;45$ parts; the maximum distance at quadrature is $R - e + r = 44;37$ parts; and the minimum distance at quadrature is $R - e - r = 34;7$ parts (all relative distances). See also Pedersen, *Survey*, 184–87; Swerdlow, *Ptolemy's Theory*, 35–37.

4. *Almagest*, v, 13; Manitius, 1:299–301.

5. *Almagest*, v, 13; Manitius, 1:301–5. See also Neugebauer's discussion of Ptolemy's errors in his determination of the Moon's distance, *HAMA*, 101–3.

6. As late as the thirteenth century, al-Tūsī still accepted the absolute distances that follow from Ptolemy's lunar model. See Hartner, "Nasir al-Dīn al-Tūsī's Lunar Theory," 291–92. In the fourteenth century Ibn al-Shātīr and Levi ben Gerson both drew attention to this discrepancy. See Goldstein, "Theory and Observation," 46–47.

7. *Almagest*, v, 14; Manitius, 1:305–6.

8. *Almagest*, v, 14; Manitius, 1:306–9. See also Swerdlow, *Ptolemy's Theory*, 52–58.

9. *Almagest*, v, 14; Manitius, 1:310–12. Ptolemy used chords instead of sines. See also Swerdlow, *Ptolemy's Theory*, 58–63.

10. Swerdlow, *Ptolemy's Theory*, 63–69.

11. *HAMA*, 106–7.

12. Hartner, "Ptolemy and Ibn Yūnus on Solar Parallax," 26.

13. R. R. Newton, *The Crime of Claudius Ptolemy*, 186–204.

14. *JKGW*, 8:414. I have used the translation by Drake and O'Malley in *Controversy*, 341.

15. Kepler, "Hipparchus," in *JKAOO*, 3:520–21.
16. *Almagest*, v, 18; Manitius, 1:323.
17. *Almagest*, ix, 1; Manitius, 2:93. See also Goldstein, "Theory and Observation," 42–44; idem, "Some Medieval Reports."
18. *Almagest*, ix, 1; Manitius, 2:93–94. See also chap. 2, note 22.
19. E.g., Kuhn, *The Copernican Revolution*, 80; *HOA*, 188–89.
20. Hartner, "Medieval Views." See also *Arabic Version*, 5–9.
21. *Arabic Version*, 6–7.
22. *Almagest*, x, 2; Manitius, 2:158–61.
23. *Arabic Version*, 7.
24. Ibid.
25. Ibid.
26. Proclus used this ratio in his *Hypotyposis*; see *HAMA*, 920, table 24, col. 4. In Proclus, *Hypotyposis*, 223–25, the ratio is erroneously given as 91;30:33;15 (sexagesimal numbers). Though Manitius, the editor of *Hypotyposis*, knew that 33;15 was wrong, he could not derive the correct number (p. 307). See also Hartner, "Medieval Views"; Goldstein and Swerdlow, "Anonymous Arabic Treatise," 156–57.
27. Al-Farghānī and al-Battānī did just that; see chap. 4.
28. *Arabic Version*, 7.
29. Ibid.
30. Ibid.
31. Ibid., 8.
32. Ibid.
33. *Almagest*, v, 16; Manitius, 1:313. The Sun's absolute radius is 1,210 (sin 15′40″) = 5½ e.r.
34. *Arabic Version*, 8–9.
35. Ibid., 9.
36. In Lindberg, *Science in the Middle Ages*, cosmology and astronomy are the subjects of different chapters written by different authors, Grant and Pedersen. Each chapter, however, has a section on physical cosmology vs. mathematical astronomy (280–84, 320–22). Ptolemy's *Planetary Hypotheses* figures prominently in both sections.
37. *HAMA*, 942–43.
38. Proclus, *In Platonis Timaeus*, 3:62–63; idem, *Hypotyposis*, 221.
39. Simplicius, *In Aristotelis de Caelo Commentaria*, 474.
40. *HAMA*, 918–19.

Chapter Four

1. Lewis, *The Discarded Image*, 13.
2. Ibid., 10.
3. Ibid., 11.
4. Ibid., 12.
5. Ibid., 97.
6. Ibid., 97–98. Note that for Maimonides and Bacon, Lewis cites Lovejoy's *Great Chain of Being*, 100.
7. Lewis, *The Discarded Image*, 22–91.
8. Pedersen, *Survey*, 14–15.
9. *Arabic Version*, 5.

10. Al-Farghānī, *Differentie*, 38–40 (chaps. 21–22).
11. *HAMA*, 163–64; Pedersen, *Survey*, 319–24.
12. Swerdlow, *Ptolemy's Theory*, 138–40.
13. Al-Farghānī, *Differentie*, 38–39 (chap. 21).
14. Ibid., 13–14 (chap. 8). According to this measure the Earth's circumference is 360 × 56⅔, or 20,400 miles, and its radius is 3,246.76 miles, rounded off to 3,250 miles by al-Farghani. For the actual measurement, see al-Bīrūnī, *The Determination of the Coordinates*, 177–80.
15. Thābit, *De Hiis que Indigent Expositione antequam Legatur Almagesti*, 136–37; Swerdlow, *Ptolemy's Theory*, 141, 175–76.
16. Al-Battānī, *Opus Astronomicum*, 1:120.
17. Ibid., 1:60–61. See also Swerdlow, "Al-Battānī's Determination of Solar Distance."
18. Al-Battānī, *Opus Astronomicum*, 1:120–24.
19. Swerdlow, *Ptolemy's Theory*, 148–56, 182–87.
20. Saliba, "The First Non-Ptolemaic Astronomy at the Maraghah School."
21. Goldstein and Swerdlow, "Anonymous Arabic Treatise," 148.
22. Ibid., 148–53.
23. Pedersen, "Astronomy," 313.
24. *HOA*, 188–89, 257–58; Kuhn, *The Copernican Revolution*, 80–81.
25. John of Seville's translation of 1137, *Differentie*, was much more popular than Gerard of Cremona's translation of 1175, *Liber de Aggregationibus*. In Oxford alone, Paget Toynbee found twenty manuscript copies of al-Farghānī's tract; seventeen are of John's translation. See Toynbee, "Dante's Obligation," 413–15. John's translation was printed in numerous editions, beginning in 1493. The title *Elementa Astronomica*, often used today, was introduced in the edition of Jacob Christmann (Frankfurt, 1590) and used by Jacob Golius in his retranslation from the Arabic (Amsterdam, 1669). Gerard's translation was not printed until 1910.
26. Benjamin and Toomer, *Campanus of Novara*, 382 n. 20. See also n. 14 above.
27. Ibid., 343, 363.
28. Thorndike, *Sphere of Sacrobosco*, 195, 243. The numbers given here suffer severely from scribal errors. See also ibid., 469 n. 14.
29. Benjamin and Toomer list sixty-one copies of Campanus's *Theorica Planetarum* (*Campanus of Novara*, 59–120). Thorndike lists thirteen copies of Robertus's commentary on Sacrobosco's *Sphere*, but states that this list is "probably far from being complete" (*Sphere of Sacrobosco*, 59–60, 72–75).
30. Carmody, "Leopold of Austria 'Li Compilacions de le Science des Estoilles,'" 69.
31. Carmody, *The Astronomical Works of Thabit b. Qurra*, 137, 146.
32. E.g., *Summa Philosophiae*, probably written by Robert Grosseteste, gives the distances of al-Farghānī, although the distances of the superior planets are corrupted; see Grosseteste, *Die Philosophischen Werke des Robert Grosseteste*, 9: 572–75. Pierre d'Ailly gives the sizes of the heavenly bodies according to al-Farghānī in his *Ymago Mundi*, 634.
33. Bacon, *Opus Maius*, 1: 249–50.
34. Lovejoy, *The Great Chain of Being*, 100.
35. Bacon, *Opus Maius*, 1:250.
36. Ibid., 1:250–51.

37. Ibid., 1:258–59.

38. Ibid., 1:258.

39. *Caxton's Mirrour*, 170–71. The same information is found in Brunetto Latini's *Li Livres dou Tresor*, composed around 1265–68. See Carmody, "Li Livres dou Tresor de Brunetto Latini," 100–101.

40. *Caxton's Mirrour*, 171–72.

41. Ibid., 172.

42. Ibid., 125–26, 169.

43. Orosius, in his *Historiarum adversus Paganos Libri VII*, written in 418 A.D., gives two dates for Adam's creation: 5199 and 5200 B.C. (pp. 6, 564).

44. D'Evelyn and Mill, *The South English Legendary*, 2:415.

45. Ibid., 417–18.

46. Al-Farghānī, *Differentie*, 13–14.

47. Maimonides, *The Guide of the Perplexed*, 456.

48. Bacon, *Opus Maius*, 1:247. Bacon also assumes (1:252) that the Gallic mile equals two English miles. Apparently al-Farghānī's mile was confused with the Gallic mile by the author of *The South English Legendary*.

49. *Caxton's Mirrour*, 171.

50. Dante, *Convivio*, iv, 8. I quote here from Orr, *Dante and the Early Astronomers*, 309. See also Toynbee, "Dante's Obligation," 432.

51. Dante, *Convivio*, ii, 14; Toynbee, "Dante's Obligation," 423. The translation is my own.

52. Dante, *Convivio*, ii, 7; Toynbee, "Dante's Obligation," 422.

53. Dante, *The Divine Comedy*, vol. 3, *Paradiso*, part 1, 103.

54. Al-Farghānī, *Differentie*, 38.

55. Orr, *Dante and the Early Astronomers*, 310–11.

56. Dreyer, "Mediaeval Astronomy," in *Studies in the History and Method of Science*, 2:116, and in *Toward Modern Science*, 1:251.

57. Zinner, *Entstehung und Ausbreitung*, 81–84. See also Kren, "Homocentric Astronomy in the Latin West."

58. Goldstein, *The Astronomical Tables of Levi ben Gerson*, 28–29.

Chapter Five

1. Hellman and Swerdlow, "Peurbach," 473–79; Rosen, "Regiomontanus," 348–52. Regiomontanus lectured on al-Farghānī at Padua in 1464 and made additions to and notes on al-Battānī's *De Motu Stellarum*. Al-Farghānī's *Differentie* (trans. John of Seville), Regiomontanus's lecture on it, al-Battānī's *De Motu Stellarum*, and Regiomontanus's notes to it were printed together in 1537 under the title *Continentur in hoc Libro Rudimenta Astronomica Alfragani &c*. See also *HOA*, 281; Kuhn, *The Copernican Revolution*, 122–32. Note that Copernicus's *De Revolutionibus* does not contain a single reference to al-Farghānī.

2. Regiomontanus, *Epytoma Joannis de monte regio In Almagestum Ptolomei*, v, 19–20; fols. f3v–f5r. Note that Peurbach did criticize Ptolemy's lunar theory because it predicted a manifestly erroneous variation in the Moon's apparent diameter of almost a factor of 2 (v, 21; fol. f5r).

3. Ibid., ix, 1; fol. k1v.

4. *De Rev.*, i, 10; *NCR*, 18–22.

5. *Almagest*, x, 2; Manitius, 2:158–61.

6. *De Rev.*, v, 21; *NCR*, 271. See also *NCR*, 427 n. P.271:39.
7. *De Rev.*, v, 27; *NCR*, 281.
8. *De Rev.*, v, 9, 14, 19; *NCR*, 252–54, 261–62, 268–70.
9. *HAMA*, 146.
10. *De Rev.*, v, 9, 14, 19, 21, 27; *NCR*, 254, 262, 270, 271, 281.
11. *De Rev.*, iv, 16–17; *NCR*, 203–5. Note that Copernicus, too, criticized Ptolemy's lunar theory for wrongly predicting a very large variation in the Moon's apparent diameter, see *De Rev.*, iv, 2; *NCR*, 176.
12. *De Rev.*, iv, 19; *NCR*, 206–7.
13. *De Rev.*, iv, 19; *NCR*, 207.
14. Copernicus found the Sun's eccentricity to be 1 part in 31, compared to Ptolemy's 1 part in 24. He concluded that the solar eccentricity varied with a definite period and constructed a model which produced a maximum eccentricity in 64 B.C. and a minimum in the seventeenth century. See *HOA*, 331–32. Note that while the solar apogee distance had thus decreased, according to Copernicus, its perigee distance had increased.
15. *De Rev.*, iv, 19; *NCR*, 207.
16. Henderson, *On the Distances*, 156–65. See also idem, "Erasmus Reinhold's Determination;" Abers and Kennel, "Commentary: The Role of Error in Ancient Methods," 130.
17. This inevitably raises the question of what Copernicus meant by the word "sphere." See Rosen, "Copernicus' Spheres and Epicycles;" Swerdlow, "Pseudodoxia Copernicana;" Rosen, "Reply to N. Swerdlow."
18. *De Rev.*, i, 10; *NCR*, 19.
19. *De Rev.*, prefatory letter to Pope Paul III; *NCR*, 5.
20. Henderson, *On the Distances*, 137–90; idem, "Erasmus Reinhold's Determination," 124–28.
21. For Michael Maestlin's redetermination of Reinhold's results and its use by Christoph Scheiner and Galileo, see chaps. 5 and 6.
22. According to Copernicus, Mars's least distance from the Sun was 1.3739 a.u., and the Earth's greatest distance from the Sun was 1.0322 a.u. With Mars at opposition and near its perihelion, and with the Earth near aphelion, the Earth-Mars distance was, therefore, near a minimum of $1.3739 - 1.0322 = 0.3417$ a.u. Assuming a horizontal solar parallax at mean solar distance of $3'$, Mars's parallax should be almost $9'$ at such favorable oppositions.
23. Tycho to Henricus Brucaeus, 1584, *TBOO*, 7:80. As Dreyer pointed out, at this time Tycho was only trying to prove the Copernican theory wrong. See Dreyer, *Tycho Brahe*, 179; *HOA*, 363.
24. Tycho to the Landgraf, 18 January 1587, *TBOO*, 6:70; Tycho to Caspar Peucer, 13 September 1588, *TBOO*, 7:129; Tycho to Christoph Rothmann, 21 February 1589, *TBOO*, 6:179.
25. See Dreyer's "Prolegomena Editoris," *TBOO*, 1:xxix–xl; Tycho to Peucer, *TBOO*, 7:129; Gingerich, "Dreyer and Tycho's World System."
26. Tycho, *Astronomiae Reformatae Progymnasmata* (1602), *TBOO*, 2:421–22. Tycho mentioned Johannes Francus Offusius, who based his celestial distances on harmonies and made the solar distance 576 terrestrial diameters. See also Dreyer's note, *TBOO*, 460.
27. *TBOO*, 2:428–30. Note that the diameter of the "sphere" that would entirely contain the deferent and both epicycles of Saturn was 12,900 e.r. At apogee,

however, Saturn was at the bottom of its second epicycle, not the top, according to Tycho's and Copernicus's model of the planet. Since the radius of this second epicycle was 300 e.r., Saturn's apogee distance was 12,300 e.r.

28. *TBOO*, 2:422–26, 430–31. Tycho introduced a subdivision, here ignored, in first magnitude fixed stars. Some of these, e.g., Sirius and Lyra, were somewhat larger than ordinary first magnitude stars. Their apparent diameter was about 2¼′ and their volume, compared to the Earth's, almost 100. Likewise, some first magnitude stars were somewhat smaller than ordinary ones: their apparent diameter was about 1¾′ and their volume not much more than 45 (*TBOO*, 2:431). Note that the section headings indicate that all the figures for sizes were "aestimationes," "opiniones," and "censurae."

29. *De Rev.*, i, 10; *NCR*, 22.

30. Tycho to Rothmann, 24 November 1589, *TBOO*, 6:197.

31. See n. 29.

32. *TBOO*, 6:197.

33. Rothmann to Tycho, 18 April 1590, *TBOO*, 6:216.

34. Magini, *Novae Coelestium Orbium Theorica*, 102v–103r.

35. Ibid., 107v–109r. Magini described the eclipse method for determining solar distance and gave the distances found by Ptolemy and Copernicus.

36. Clavius, *In Sphaeram Ioannis de Sacro Bosco Commentarium*, in *Opera Mathematica*, 3:100–102, 117.

Chapter Six

1. Kepler, *Prodromus Dissertationum Cosmographicum, Continens Mysterium Cosmographicum* (1597), in *JKGW*, 1:12. I have used the translation in Duncan, 67.

2. Plato used the five regular solids in his matter theory in *Timaeus*, 53c–55c. The proof that there can be only five regular (convex) polyhedra is probably the work of Theaetetus (ca. 415–369 B.C.), a member of the Platonic school. Kepler refers the reader to the proof in Euclid's *Elements*, book 13, prop. 18. See *JKGW*, 1:13; Duncan, 67.

3. *JKGW*, 1:46; Duncan, 152. For the precise derivation of these numbers, see *JKGW*, 1:425 n. 46:28.

4. Kepler to Maestlin, 3 October 1595 (o.s.), *JKGW*, 13:44. The maximum distance of Jupiter, given in sexagesimal numbers, 5. 21. 29, is a mistake for 5. 27. 29; see *De Rev.*, v, 14.

5. These are the ratios calculated from the numbers above. Kepler's ratios varied somewhat from these numbers, see below.

6. *JKGW*, 1:48; Duncan, 157.

7. *JKGW*, 1:48–50; Duncan, 157.

8. Maestlin to Kepler, 27 February 1596 (o.s.), *JKGW*, 13:55. In the following discussion I have drawn heavily on Grafton, "Michael Maestlin's Account," see especially p. 525.

9. *JKGW*, 13:56–65.

10. Grafton, "Michael Maestlin's Account," 529–31.

11. *JKGW*, 12:55; Grafton, "Michael Maestlin's Account," 525.

12. *JKGW*, 12:72; Grafton, "Michael Maestlin's Account," 531.

13. *JKGW*, 1:56–57, 76–77; Duncan, 165–67, 215–19.

14. *JKGW*, 1:52–53; Duncan, 161–62, 230–31, 244 n. 3. Maestlin actually produced two figures, one for the time of Ptolemy, ca. 140 A.D., and one for the time of Copernicus, ca. 1525.

15. *JKGW*, 1:54; Duncan, 163. For an explanation of how Kepler and Maestlin arrived at their distances of Venus and Mercury, see the notes by Caspar (*JKGW*, 1:426–29) and Aiton (Duncan, 244–46 n. 4). See also Grafton, "Michael Maestlin's Account," 529–31.

16. *JKGW*, 1:69–70; Duncan, 199.

17. *JKGW*, 1:70; Duncan, 199–201.

18. *JKGW*, 1:70–71; Duncan, 201.

19. *JKGW*, 1:71; Duncan, 201.

20. *JKGW*, 1:133–45; Grafton, "Michael Maestlin's Account," 532–50.

21. *JKGW*, 1:135–37; Grafton, "Michael Maestlin's Account," 536–39.

22. Caspar, *Kepler*, 139–46.

23. *JKGW*, 2:265–78.

24. *JKGW*, 3:120–21.

25. *TBOO*, 1:xxxix–xl. See also Gingerich, "Dreyer and Tycho's World System."

26. *JKGW*, 3:129.

27. Ibid.

28. Ibid.

29. *JKAOO*, 3:520–49.

30. Ibid., 520–21.

31. Ibid., 550–643.

32. Kepler, *De Stella Nova in Pede Serpentarii* (1606), in *JKGW*, 1:234.

33. *JKGW*, 1:235. I have corrected the number in the text, 34,077,066⅔.

34. *JKGW*, 1:238.

35. *JKGW*, 1:253. I have used the translation by Koyré, *Closed World*, 60.

36. *JKGW*, 1:253–54. Kepler's "Etenim sumamus, exempli causa, tres stellas secundae magnitudinis in cingulo Orionis," is translated by Koyré: "Indeed, let us take, for instance, three stars of the second magnitude in the belt of Orion" (*Closed World*, 63). Kepler obviously means "*the* three stars of the second magnitude," for there are only three second-magnitude stars in the belt. They are, from east to west, Zeta, Epsilon, and Delta Orionis, also known by their Arabic names, Mintaka, Alnilam, and Alnitak. Their positions (in ecliptic coordinates) were given by Tycho Brahe as follows (*TBOO*, 2:277):

	longitude Gemini	latitude
Prima baltei	16°50½′	23°38′
Media baltei	17°54′	24°33½′
Ultima baltei	19°6½′	25°21½′

Note that Tycho Brahe estimated the diameter of a second-magnitude star to be only 1½′ (table 6, p. 50), not 2′.

37. *JKGW*, 1:254; Koyré, *Closed World*, 63.

38. *JKGW*, 1:254–57; Koyré, *Closed World*, 63–72.

39. *JKGW*, 2: 264. The original passage can be found in Einhard, *Annales Laurissensis et Einhardi*, 194. See also Sarton, "Early Observations of the Sunspots?" Goldstein, "Some Medieval Reports."

40. *JKGW*, 2: 264–65.

41. Maestlin, *Disputatio in Multivariis Motuum Planetarum in Coelo apparentibus Irregularitatibus, seu regularibus Inaequalitatibus, earumque Causis Astronomicis. Quam praeside Michaele Maestlino . . . defendere conabitur . . . Samuel Hafenreffer* (Tübingen, 1606), thesis 98. This work is now lost. See *JKGW*, 4:489; Rosen, *Kepler's Somnium*, 141–42 n. 385. See also Kepler to Hafenreffer, 16 November 1606, *JKGW*, 15:359–62.

42. Kepler, *Phaenomenon Singulare seu Mercurius in Sole* (1609), in *JKGW*, 4: 83–87, 92–95.

Chapter Seven

1. *Ambassades du Roy de Siam*, 9–11. The quotation is on p. 11. The tract was reprinted in Lyons in November 1608. Both editions are extremely rare. For a photographic reprint of the first edition, see Drake, *The Unsung Journalist*. See also Van Helden, *The Invention of the Telescope*, 40–42.

2. Van Helden, *The Invention of the Telescope*, 25–28. North, "Thomas Harriot and the First Telescopic Observations of Sunspots"; Rosen, "When Did Galileo Make His First Telescope?" Drake, "Galileo's First Telescopes"; idem, *Galileo at Work*, 137–61; idem, "Galileo's First Telescopic Observations."

3. Galileo, *Sidereus Nuncius* (1610), in *OGG*, vol. 3, part 1, 75; *Discoveries*, 45–46.

4. *OGG*, 1:52, 54; Wallace, *Galileo's Early Notebooks*, 76, 79.

5. *OGG*, vol. 3, part 1, 75; *Discoveries*, 46.

6. *OGG*, vol. 3, part 1, 75–76; *Discoveries*, 46.

7. *JKGW*, 4:203; Rosen, *Kepler's Conversation*, 33–34.

8. *OGG*, vol. 3, part 1, 76; *Discoveries*, 47.

9. *OGG*, vol. 3, part 1, 61–62; *Discoveries*, 30–31.

10. *JKGW*, 4:322.

11. Christmann, *Nodus Gordius*, 41.

12. *JKGW*, 4:346–49; Carlos, *The Sidereal Messenger of Galileo*, 93–102.

13. Galileo to Sarpi, 12 February 1612, *OGG*, 11:49.

14. "Apelles hiding behind the painting." See *Discoveries*, 82 n. 19.

15. Scheiner, *Tres Epistolae de Maculis Solaribus*, in *OGG*, 5:28. This letter is dated 19 December 1611.

16. Galileo, *Istoria e Dimostrazioni intorno alle Macchie Solari e loro Accidenti* (1613), in *OGG*, 5:100; *Discoveries*, 94. This first letter was dated 4 May 1612.

17. Scheiner, *Accuratior Disquisitio*, in *OGG*, 5:41. This section of the tract is dated 16 January 1612.

18. *OGG*, 5:196–97; *Discoveries*, 131–32. This is Galileo's third letter, dated 1 December 1612.

19. Galileo, *Lettera a Madama Cristina di Lorena Granduchessa di Toscana* (1615), in *OGG*, 5:328; *Discoveries*, 195–96. Idem, *Il Saggiatore* (1623), in *OGG*, 6:232–33; *Controversy*, 184.

20. *OGG*, 5:197; *Discoveries*, 132.

21. Locher, *Disquisitiones Mathematicae*, 25–28.

22. Marius, *Mundus Iovialis* (1614); Prickard, "The 'Mundus Jovialis' of Simon Marius," 374.

23. *JKGW*, 7:282.

24. *OGG*, 6:525.

25. Ibid., 524–26.

26. *JKGW*, 8:417; *Controversy*, 344.

27. Locher, *Disquisitiones Mathematicae*, 26. Scheiner and Locher calculated that a first magnitude star has a diameter $3\frac{181}{604}$ and a third magnitude star a diameter $1\frac{785}{1{,}208}$ times the Earth's annual orbit, while a sixth magnitude fixed star "does not leave much of the diameter of the annual orb."

28. *OGG*, 7:386; *Dialogue*, 359.

29. *OGG*, 6:525. This statement is explained as follows:

(1) The traditional volume of Jupiter was about 80 times the Earth's, based on an apparent diameter of $\frac{1}{12}$th the Sun's, or $2\frac{1}{2}'$. An error of $2\frac{1}{2}':40'' =$ ca. 4 would mean an error in volume of 4^3, or about 64, not a factor of 2,400.

(2) Magini had made Jupiter's apparent diameter at perigee to be about $9'$ (see p. 52). This entailed an error of $9':40'' =$ ca. 13.5, or an error in volume of 13.5^3 = ca. 2,400.

(3) Confusing Magini's apparent diameter with the traditional one, Galileo now argued that Jupiter's volume was not 80 times greater, but rather $\frac{80}{2{,}400}$, or 30 times smaller than the Earth's.

30. *OGG*, 7:386–87; *Dialogue*, 359–60.

31. Koyré, *Closed World*, 95–99.

32. *OGG*, 7:387–88; *Dialogue*, 360–61. I have changed the spelling of the names of the Moslem astronomers.

33. *OGG*, 7:388; *Dialogue*, 361.

34. *OGG*, 7:388–89; *Dialogue*, 361.

35. *OGG*, 7:388–92; *Dialogue*, 361–64.

Chapter Eight

1. *JKGW*, 4:295; Rosen, *Kepler's Conversation*, 22.

2. *JKGW*, 4:303; Rosen, *Kepler's Conversation*, 35.

3. The telescope had, of course, reached Prague before March 1610. On 19 April Giuliano de' Medici asked Galileo to send one of his telescopes to Prague since the ones available there were not good enough to show what Galileo wrote about in his *Sidereus Nuncius*. See *JKGW*, 16:303; *OGG*, 10:319.

4. *JKAOO*, 7:482–83.

5. Ibid., 483.

6. Ibid.

7. Ibid., 486.

8. Shirley, "Thomas Harriot's Lunar Observations," 304–6.

9. *JKAOO*, 7:516.

10. See also Kepler's statement cited on p. 19; *JKGW*, 8:414; *Controversy*, 341.

11. *JKAOO*, 7:496.

12. Ibid., 528.

13. Ibid.

14. Kepler, *Harmonice Mundi* (1619), in *JKGW*, 6:309.

15. *JKGW*, 6:302. See also Gingerich, "The Origins of Kepler's Third Law."

16. *JKGW*, 6:358. See also *HOA*, 405–10.

17. Kepler, *Epitome Astronomiae Copernicanae* (1618–21), in *JKGW*, 7:249. *Epitome*, 845.

18. *JKGW*, 7:264–65; *Epitome*, 861.

19. *JKGW*, 7:276–77; *Epitome*, 873.
20. *JKGW*, 7:277–78; *Epitome*, 874–75.
21. *JKGW*, 7:278; *Epitome*, 875. I have made changes in Wallis's translation.
22. *JKGW*, 7:278–79; *Epitome*, 875.
23. *JKGW*, 7:279–81; *Epitome*, 876–78.
24. *JKGW*, 7:281; *Epitome*, 877.
25. *JKGW*, 7:281; *Epitome*, 878. I have made changes in Wallis's translation.
26. *JKGW*, 6:308.
27. *JKGW*, 7:282; *Epitome*, 879. I have made changes in Wallis's translation.
28. *JKGW*, 7:282; *Epitome*, 879. I have made changes in Wallis's translation.
29. *JKGW*, 7:282. I have taken the translation from Koyré and Madison, *The Astronomical Revolution*, 351–52.
30. *JKGW*, 1:71; 3:236–42.
31. *JKGW*, 7:306–7; *Epitome*, 905. I have made changes in Wallis's translation.
32. *JKGW*, 7:307.
33. Ibid., 283–84. See also Gingerich, "The Origins of Kepler's Third Law;" Hoyer, "Über die Unvereinbarkeit," 196–97.
34. *JKGW*, 7:284.
35. Ibid., 285.
36. Ibid., 285–86.
37. Ibid., 286.
38. Ibid., 286–87.
39. Ibid., 287.
40. Ibid., 1:9–10; Duncan, 63.
41. *JKGW*, 7:288.
42. Ibid.
43. It seems clear that Kepler ignored the volume of the Sun and took the cube root of $64 \times 10^{18} + 8 \times 10^{9}$ by using the first two terms of the infinite series:

$$(1 + e)^{-1/3} = 1 + \tfrac{1}{3} e - \tfrac{1}{9} e^{2} \ldots$$

$$\text{Thus, } 4 \times 10^{6} \left(1 + \frac{8 \times 10^{9}}{64 \times 10^{18}}\right)$$

$$= 4 \times 10^{6} \left(1 + \tfrac{1}{3} \times \frac{8 \times 10^{9}}{64 \times 10^{18}}\right)$$

$$= 4 \times 10^{6} + 6 \times 10^{-3}.$$

44. *JKGW*, 1:136. This was a fairly standard value, based on 15 German miles per degree at the equator. This would make the German mile about 4.4 English miles.
45. Ibid., 7:288: "cum jam non sit crassa magis, quam unius semid: corporis Solaris sexies millesimam vel duo milliaria Germanica, plus." Wallis (*Epitome*, 886) and Koyré, followed by Madison (*Astronomical Revolution*, 361), translate *duo milliaria Germanica* as *2,000* German miles!
46. *JKGW*, 7:288–89; *Epitome*, 886–87.
47. *JKGW*, 7:289; *Epitome*, 886.
48. *HAMA*, 892–96.
49. On Walther, see Beaver, "Bernard Walther: Innovator in Astronomical Observation," and Kremer, "The Use of Walther's Astronomical Observations." For Tycho's investigations of refraction, see, e.g., *TBOO*, 11:324–25; 12:270, 313, 373.

50. Tycho, *Astronomiae Instauratae Progymnasmata* (1602), in *TBOO*, 2:76. Risner, *Opticae Thesaurus*.

51. Risner, *Opticae Thesaurus*, 251–52, 444–45. Tycho was skeptical about their claim. See *TBOO*, 2:76.

52. *TBOO*, 2:76–85, 130–36.

53. Ibid., 64, 136.

54. Ibid., 286–88.

55. Ibid., 64.

56. Ibid., 77, 445n; ibid., vol. 6.

57. Ibid., 2:76–77.

58. Ibid., 286.

59. *JKGW*, 2:106.

60. E.g., *TBOO*, 2:85–86. For the relationship between solar theory and the refraction and parallax corrections, see Maeyama, "The Historical Development of Solar Theories."

61. Dreyer, *Tycho Brahe*, 123.

62. *JKGW*, 2:104–7.

63. Ibid., 117. See also Buchdahl, "Methodological Aspects of Kepler's Theory of Refraction."

64. *JKGW*, 10:242–43, and 142 of the tables.

65. Remus to Kepler, 4 October 1619, *JKGW*, 17:388.

66. Kepler to Remus, October 1619, *JKGW*, 17:406.

67. Remus to Kepler, 11 October 1619, *JKGW*, 17:394.

68. Remus to Kepler, 11 December 1628, *JKGW*, 18:366.

69. Ibid., 366–67.

70. Ibid., 367.

71. Ibid.

72. Ibid., 367–68.

73. *JKGW*, 8:414–15; *Controversy*, 341.

74. *JKGW*, 18:388.

Chapter Nine

1. *OGG*, 7:367; *Dialogue*, 339. It is not known when the phases of Mercury were first observed or by whom. Johannes Hevelius claimed to have observed more or less the complete cycle between November 1644 and May 1645 (*Selenographia*, 74–76, and fig. K opposite p. 70). Francesco Fontana claimed to have observed Mercury's crescent in January 1646 and reported a similar observation made by Giovanni Battista Zupo in 1639 (Fontana, *Novae Coelestium*, 89–90).

2. *OGG*, 5:98–100; *Discoveries*, 93–95.

3. Caspar, *Bibliographia Kepleriana*, 92, 92, and the facsimiles of the title pages, 81, 83. The text is in *JKAOO*, 7:589–94. See also Van Helden, "The Importance of the Transit of Mercury of 1631."

4. *JKAOO*, 7:593–94.

5. Ibid., 592–93.

6. Ibid., 593; *JKGW*, 10:166–67.

7. *JKAOO*, 7:594.

8. Gassendi, *Mercurius in Sole Visa et Venus Invisa* (1632), in *PGOO*, 4:500.

9. Ibid.

10. Ibid., 500–501.

11. Ibid., 501.
12. Ibid., 502.
13. Barrettus, *Historia Coelestis*, 955. Cysat, Letter to Johannes Lanz, 378.
14. Barrettus, *Historia Coelestis*, 955–56. See also Pingré, *Annales Célestes*, 84.
15. Barrettus, *Historia Coelestis*, 956.
16. Galileo to Elia Diodati, 15 January 1633, *OGG*, 15:26.
17. Peiresc to Gassendi, 22 December 1631, cited in Humbert, "A Propos du Passage de Mercure 1631," 31.
18. Schickard, *Pars Responsi*, 14.
19. Ibid., 4.
20. Ibid., 9.
21. Ibid., 12–14.
22. Colombe, *Contro il Moto della Terra* (1610), in *OGG*, 3:286–87.
23. Schickard, *Pars Responsi*, 14.
24. Hortensius, *Dissertatio de Mercurio in Sole Viso*, 7.
25. Ibid., 23–29.
26. Ibid., 51–53.
27. Ibid., 53–55.
28. Ibid, 61; see also 55–59.
29. Ibid., 60–63.
30. *PGOO*, 4:501.

Chapter Ten

1. *JKAOO*, 7:592.
2. The periodicity and rarity of Venus transits is shown by their occurrences between 1518 and 2012:

2 June	1518	1 June	1526
7 December	1631	4 December	1639
6 June	1761	3 June	1769
9 December	1874	6 December	1882
8 June	2004	6 June	2012

3. Lansberge, *Commentationes in Motum Terrae Diurnum, et Annuum*.
4. Horrocks, *Venus in Sole Visa*, 111; Whatton, 111.
5. Wilson, "Horrocks, Harmonies, and the Exactitude of Kepler's Third Law," 248–50.
6. *Almagest*, iii, 4; Manitius, 1:170. *De Rev.*, iii, 17; *NCR*, 159.
7. Tycho, *Astronomiae Instauratae Progymnasmata*, *TBOO*, 2:21.
8. *JHOP*, 301.
9. Ibid., 102–67.
10. E.g., ibid., 333.
11. Ibid., 331.
12. Ibid., 307.
13. Ibid., 330. In *Venus in Sole Visa*, Horrocks makes no mention of Hortensius's *Dissertatio de Mercurio in Sole Viso*.
14. Horrocks, *Venus in Sole Visa*, 115–19; Whatton, 123–29. See also Gaythorpe, "Horrocks's Observation of the Transit of Venus," parts 1 and 2.
15. *JHOP*, 330.

16. Ibid., 395–96. The problem of the accuracy of these measurements is compounded by the many errors in *JHOP*. Thus, on 2 March 1640 (o.s.) Horrocks made the following entry (ibid., 396): "Jupiter & Venus aequales. Jupiter ad distantiam 4200, tegebatur ab acu 8. Ergo diameter ejus minor scr. 0′39″." But sin 39″ = 8/42,000, not 8/4,200.

17. Ibid., 338.

18. Applebaum, "Jeremiah Horrocks," 516. *Venus in Sole Visa* was published by Hevelius in 1662 as an appendix to his *Mercurius in Sole Visus*, and in 1673 John Wallis printed Horrocks's other astronomical writings in *Opera Posthuma*, reprinted in 1678. For new information on the circulation of Horrocks's manuscripts in the late 1650s, see Applebaum and Hatch, "Boulliau, Mercator, and Horrocks's *Venus in Sole Visa*."

19. Horrocks, *Venus in Sole Visa*, 117; Whatton, 130.

20. Horrocks, *Venus in Sole Visa*, 137; Whatton, 187.

21. Horrocks, *Venus in Sole Visa*, 137–39; Whatton, 190–201.

22. Horrocks, *Venus in Sole Visa*, 141; Whatton, 202. For Horrocks's critique of the eclipse method and Ptolemy's and Lansberge's use of it, see *JHOP*, 102–67.

23. Horrocks, *Venus in Sole Visa*, 141; Whatton, 202–5.

24. Horrocks, *Venus in Sole Visa*, 141–42; Whatton, 206–7.

25. Horrocks, *Venus in Sole Visa*, 142; Whatton, 207.

26. Horrocks, *Venus in Sole Visa*, 142; Whatton, 209–11.

27. Horrocks, *Venus in Sole Visa*, 142; Whatton, 212.

28. Horrocks, *Venus in Sole Visa*, 143; Whatton, 213–15.

29. *Epitome*, 879.

30. Crabtree to Horrocks, 1639, cited in "A Catalogue of the Most Eminent Astronomers, Ancient to Modern," entry 92, "William Gascoigne." This catalog is appended to Sherburne, *The Sphere of Marcus Manilius*.

31. Horrocks, *Venus in Sole Visa*, 141; Whatton, 209–11.

32. On 5 June 1638 (o.s.) he wrote to Crabtree (*JHOP*, 309): "Mars appears of about the same size as Jupiter: but Kepler and Lansberge make him much bigger. If Mars is larger than the Earth, the solar parallax has to be much smaller than Kepler wants."

33. *JHOP*, 160, 164.

34. Ibid., 164–65.

35. *PGOO*, 4:38–40, 60–61, 64–65.

36. Hérigone, *Cursus Mathematicus*, 4:62; 5:619.

37. Rheita, *Oculus Enoch et Eliae*, part 1, 206–11.

38. Hevelius, *Selenographia*, 449, 548. Hevelius compared Jupiter in size to lunar features.

39. Wendelin, *Loxias*, 11–12.

40. Wendelin to Gassendi, 1 May 1635, *PGOO*, 6:427–29. See also Wendelin to Gassendi, 25 October 1643, ibid., 460.

41. Wendelin, *Eclipses Lunares*, 29.

42. Riccioli, *Almagestum Novum*, 1:109.

43. Ibid., 1:708.

44. Ibid., 2:106–7.

45. Ibid., 108.

46. Ibid., 108–9.

47. Ibid., 109.

48. Ibid., 2:731.
49. Ibid.
50. Ibid., 732.
51. Ibid.
52. Ibid., 686–87, 693, 708, 709.
53. Ibid., 418–19.
54. Ibid., 2:668.
55. The field of view of a "Dutch" telescope depends on the size of the pupil as well. For an excellent explanation, see North, "Thomas Harriot and the First Telescopic Observations of Sunspots," 144–45, 158–60.
56. Kepler already wrote in the first part of book 4 of his *Epitome* that the more perfect the instrument the more it represents fixed stars as mere points (*JKGW*, 7:289; *Epitome*, 886).
57. Boulliau, *Astronomia Philolaica*, 193–96.
58. *PGOO*, 4:60, 64, 80.

Chapter Eleven

1. Kepler, *Dioptrice* (1611), in *JKGW*, 4:387.
2. Galileo, *Sidereus Nuncius* (1610), in *OGG*, vol. 3, part 1, 61; *Discoveries*, 29; North, "Thomas Harriot and the First Telescopic Observations of Sunspots," 142.
3. Harriot sometimes used an instrument that magnified 30 times, but usually he employed telescopes magnifying 10 or 20 times (North, "Thomas Harriot and the First Telescopic Observation of Sunspots," 141–42). Drake notes that although Galileo made a 30-powered instrument, he can find no mention of its use (*Galileo at Work*, 148).
4. Van Helden, "The Telescope in the Seventeenth Century," 46–47.
5. Gascoigne to Oughtred, 1641, in Rigaud, *Correspondence of Scientific Men*, 1:46.
6. Ibid., 52.
7. Ibid., 58–59.
8. Ibid., 54.
9. For the development of the micrometer, see McKeon, "Les Débuts." J. A. Bennett suggests that perhaps Christopher Wren was already using a micrometer modeled on Gascoigne's in mapping the Moon in 1656 (*The Mathematical Science of Christopher Wren*, 39–40).
10. Van Helden, " 'Annulo Cingitur': the Solution of the Problem of Saturn."
11. Huygens, *Systema Saturnium* (1659), in *OCCH*, 15:348–50.
12. Ibid., 234.
13. On 11 April 1672 (o.s.) Flamsteed wrote to Oldenburg: ". . . holdeing my glasse over ye flame of a candle to smeare it as I used to doe frequently it cracked in a veine through ye middle & a bit of ye breadth of a great pins head burst from it." See *Oldenburg*, 9:3.
14. *OCCH*, 15:342–52.
15. Ibid., 342–45.
16. Ibid., 344–45.
17. Ibid., 344–47.
18. Ibid., 346–47.
19. Ibid., 348–49.

20. Ibid., 212–15.
21. See also Cohen, "Perfect Numbers in the Copernican System."
22. Boulliau to Huygens, November 1659, *OCCH*, 2:512–13.
23. Streete, *Astronomia Carolina*, 61.
24. Ibid., 12.
25. Ibid., 62.
26. *Arabic Version*, 7.
27. Streete, *Astronomia Carolina*, 118. See also Huygens to Boulliau, 13 June 1661, *OCCH*, 3:279–81; 15:70–71.
28. *OCCH*, 3:281; 15:72–73. Streete, *Astronomia Carolina*, 188–99.
29. Huygens to Hevelius, 22 August 1661, *OCCH*, 3:315.
30. Streete, *Astronomia Carolina*, 118.
31. Hevelius, *Mercurius in Sole Visus*, 80–83.
32. Ibid., 83, 96–97.
33. Ibid., 95; idem, *Selenographia*, 37.
34. Huygens to Hevelius, 25 July 1662, *OCCH*, 4:181–82.
35. Hevelius, *Mercurius in Sole Visus*, 93, 143.
36. McKeon, "Les Débuts," 246–56.
37. Observatoire de Paris, MSS. D1, 14, fols. 3–16. Registre, 1:28–30.
38. Baily, *An Account*, 29–30.
39. Flamsteed, "Flamsteed Papers," 41:118v–123v.
40. There are a few instances of almost simultaneous measurements. On 8 and 12 September 1672 (n.s.) Flamsteed measured Mars's diameter to be 33″. 8 September was the day of Mars's opposition, and Flamsteed took pains in his measurement: "Diameter pluribus & saepe repetibis Experimentis inventa." (Flamsteed, *Historia Coelestis Britannica*, 1:15.) On 19 August and 5 September of that same year, Picard measured Mars's diameter to be 26″ "tout au plus" (Le Monnier, *Histoire Céleste*, 28). Likewise, on 30 March 1673 (n.s.) Flamsteed measured Jupiter's diameter to be 48″, "but certainly not less than 47″" (*Historia Coelestis Britannica*, 1:19), and on 1 April Picard measured it to be 44″ (*Histoire Céleste*, 28). Picard's measurements were somewhat lower and closer to the modern values, but this does not, of course, prove that they were in fact more accurate.

Chapter Twelve

1. E.g., Taton, "Cassini, Gian Domenico," 103; Rosen, "Richer," 424; Herrmann, *Kosmische Weiten*, 47; Abetti, *The History of Astronomy*, 127. In *A History of Astronomy*, Pannekoek is more careful, assigning to Cassini's solar parallax "the relatively considerable uncertainty of some few seconds, ⅓ or ¼ of its amount" (p. 284).
2. E.g., Registre, 2:18–19, 28, 43–58; 3:297r; 5:226r.
3. G. D. Cassini, *La Meridiana*, 5–10.
4. G. D. Cassini, *De l'Origine et du Progrès de l'Astronomie*, 37–39; idem, *Les Élémens de l'Astronomie*, 55–57; idem, *De Solaribus Hypothesibus*.
5. G. D. Cassini, *Les Élémens de l'Astronomie*, 56–58.
6. Ibid., 60–61.
7. Ibid., 61.
8. Registre, 2:18; *OCCH*, 21:31.
9. Registre, 3:298v.
10. Ibid., 5:226r.
11. Ibid.

12. G. D. Cassini, *Les Élémens de l'Astronomie*, 59.

13. Ibid., 60.

14. Picard, *Mesure de la Terre*, in *Mem.*, vol. 7, part 1, 133–90.

15. Picard, *Voyage d'Uranibourg*, in *Mem.*, vol. 7, part 1, 191–230.

16. Registre, 2:37–58. It was suggested to make simultaneous measurements of the Sun's altitudes in Madagascar and Paris. This was the only proposal for finding the Sun's parallax (p. 45).

17. Olmsted, "The Scientific Expedition," 121–24.

18. Mims, *Colbert's West India Policy*.

19. Olmsted, "The Scientific Expedition," 120–23.

20. Richer, "Observations Astronomiques," in *Mem.*, vol. 7, part 1, 236–38.

21. Ibid., 233–36. See also Registre, 9:137r–139r.

22. G. D. Cassini explained this method in *Observations sur La Comète*, 33–61.

23. Hevelius's criticism is in *Mercurius in Sole Visus*, 85, 143. Flamsteed's defense of Horrocks can be found in Rigaud, *Correspondence of Scientific Men*, 2:87, 109, 142.

24. *Oldenburg*, 8:47.

25. Baily, *An Account*, 21–22; Rigaud, *Correspondence of Scientific Men*, 2:97; *Oldenburg*, 10:156. See also Forbes, *The Gresham Lectures*, 5–10.

26. Rigaud, *Correspondence of Scientific Men*, 2:95.

27. Baily, *An Account*, 29–30.

28. *Oldenburg*, 9:4.

29. Ibid., 159. Forbes ignores Flamsteed's own estimate of the accuracy of these measurements and claims that the micrometer "enabled [Flamsteed] to measure small angular distances in the heavens to the hitherto unobtainable accuracy of about one second of arc (1″)." See Forbes, *The Gresham Lectures*, 8.

30. *Oldenburg*, 9:178–79; *Phil. Trans.*, no. 86, 5040–42.

31. *Oldenburg*, 9:178; *Phil. Trans.*, no. 86, 5041.

32. *Oldenburg*, 9:178; *Phil. Trans.*, no. 86, 5041.

33. Baily, *An Account*, 32; Forbes, *The Gresham Lectures*, 10–11, 98.

34. Forbes, *The Gresham Lectures*, 10–18, 98.

35. *Oldenburg*, 9:327.

36. Ibid., 469–70; Rigaud, *Correspondence of Scientific Men*, 2:160. The solar distance of 26,000 e.r. in the letter to Oldenburg is an error: $1/\sin 7'' = 29{,}466$.

37. Flamsteed, *Phil. Trans.*, no. 96, 6000 (a misprint for 6100).

38. *Oldenburg*, 10:108, 110–11.

39. Le Monnier, *Histoire Céleste*, 28–29.

40. *Mem.*, 8:105–7.

41. *Oldenburg*, 10:381–82.

42. *Mem.*, 8:107.

43. Ibid.

44. Académie Royale des Sciences, *Histoire . . . (1666–1699)*, 1:168.

45. *Mem.*, 8:62–65.

46. Ibid.

47. Registre, 9:137v–138v. Cassini did not even mention these efforts in his *Élémens*.

48. Mem., 8:98.

49. Ibid., 108.

50. Ibid., 69–72.
51. Ibid., 98–105, 108–13.
52. Ibid., 100–104.
53. Ibid., 104.
54. Ibid., 105.
55. Ibid., 109.
56. Ibid., 109.
57. Ibid., 113.
58. Ibid., 113–15.
59. Ibid., 115.
60. See note 1 of this chapter.
61. Wing, *Astronomia Britannica*, 121–25, and p. 128 of the tables. For the entire controversy between Streete and Wing on this subject, see Wing, Ὀλύμπια Δώματα, 15; Streete, *An Appendix to Astronomia Carolina*, 25–28; Wing, *Examen Astronomiae Carolinae*; and Streete, *Examen Examinatum*.
62. The period of a pendulum depends only on its length and the force of gravity. Since the Earth bulges at the equator, a pendulum there is subject to a lower force of gravity. Richer found that his pendulum, carefully adjusted in Paris to have a period of 1 sec., lost 2 min. 28 sec. every 24 hours in Cayenne. In order to restore the period to 1 sec. in Cayenne, he had to shorten the length of the pendulum slightly.

Chapter Thirteen

1. Armitage, *Edmond Halley*, 20–36.
2. J. Gregory, *Optica Promota*, 130.
3. Halley, *Catalogus*, appendix "Mercurii Transitus sub Solis Disco," 1–3.
4. *Journal des Sçavans*, 20 December 1677, 340–48.
5. Halley, *Catalogus*, appendix, 1–3.
6. Ibid., 3.
7. Ibid., 3–4.
8. Ibid., 4.
9. Ibid.
10. Ibid.
11. Forbes, *The Gresham Lectures*, 20, 99.
12. Ibid., 99.
13. Ibid.
14. Ibid., 100.
15. Ibid., 101.
16. A. R. Hall, "Newton on the Calculation," 62–64. Herivel, *Background*, 192–98.
17. A. R. Hall, "Newton on the Calculation," 65, 67; Herivel, *Background*, 194, 196; Isaac Newton, *Correspondence*, 1:298, 300.
18. A. R. Hall, "Newton on the Calculation," 65–66, 67–68; Isaac Newton, *Correspondence*, 1:298; Herivel, *Background*, 194–95, 196–97.
19. A. R. Hall, "Newton on the Calculation," 62–63; Isaac Newton, *Correspondence*, 2:446.
20. Isaac Newton, *Philosophiae Naturalis Principia Mathematica*, 2:481, line 15 note.
21. Ibid., 478, line 33ff note.
22. Baily, *An Account*, 33.

23. Isaac Newton, *Philosophiae Naturalis Principia Mathematica*, 2:478, line 33ff note.

24. Ibid., 579.

25. Ibid.

26. Ibid.

27. Registre, 9:107v–108r, 128v–129r.

28. Ibid., 129r.

29. G. D. Cassini, *Observations sur la Comète*, 33.

30. Registre, 9:56r.

31. G. D. Cassini, *Observations astronomiques*, 363–64.

32. Le Monnier, *Histoire Céleste*, 29.

33. *Journal des Sçavans*, 7 December 1676, 233–36; *Phil. Trans.*, no. 136 (1677): 893–94; *Mem.*, 10:575–76. Roemer wrote that at the quadratures the distance between the Earth and Jupiter changes by at least 210 terrestrial diameters during the 42½ hours it takes Jupiter's innermost satellite to complete one revolution about the planet. At these points the rapid change in the distance between Jupiter and the Earth is almost solely due to the Earth's motion. Ignoring Jupiter's motion, then, the Earth moves 420 e.r. in 42½ hours. The circumference of its annual circle is 420/42½ × 24 × 365¼ e.r., or 86,629 e.r. From this, the solar distance, the radius of the Earth's orbit, is 13,787 e.r., corresponding to a solar parallax of 15″. See Van Helden, "Roemer's Speed of Light."

34. Lalande, *Astronomie*, 2:413.

35. Bianchini, "Nova Methodus Cassiniana," 470–78.

36. Ibid., 478.

37. Huygens to J. Gousset, 2 November 1691, *OCCH*, 10:180.

38. Fontenelle, *Entretiens*, 182–83 n. 27, 191 n. 12, 193 n. 2, 195 n. 15, 196 n. 22, 198 nn. 28 and 31, 200 n. 2, 201 nn. 3 and 4.

39. Ibid., 19–20.

40. *OCCH*, 21:410. Huygens here takes Cassini's solar parallax as 10″.

41. Ibid., 410, 477.

42. Ibid., 477.

43. Isaac Newton, *Correspondence*, 3:285–87.

44. Ibid., 247.

45. Ibid., 253.

46. Bentley, *A Confutation of Atheism from the Origin and Frame of the World, Part II*, 13. See the reprint in Bentley, *Eight Lectures on Atheism*.

47. D. Gregory, *Astronomiae Physicae & Geometricae Elementa*. I quote here from the second English edition, *The Elements of Physical and Geometrical Astronomy*, 2:571.

48. Whiston, *A New Theory of the Earth* (1696), 31–32. Whiston commented on the planetary distances: "The proportions of these Numbers are unquestionable: But the Numbers themselves only within about a fourth part under or over." In later editions (e.g., London, 1737, 34–35) he used a solar parallax of 10″ in his calculations, but he did not remove the caution: "the Sun's *Parallax* . . . on which the whole depends, is not yet accurately determin'd by Astronomers; so that no exact Number can be certainly pitch'd upon, till farther Observations put an end to our Doubts."

49. Whiston, *Praelectiones Astronomicae*. Taken from the second English edition, *Astronomical Lectures* (1728), 82–85. The numbers in the three editions (1707, 1715, 1728) are the same.

50. Halley, *Phil. Trans.* 17, no. 193 (1691): 511–22, esp. 522.

51. D. Gregory, *The Elements of Physical and Geometrical Astronomy*, 2:482–83.

52. Ibid., 571.

53. Huygens, *Cosmotheoros* (1698), in *OCCH*, 21:693, 783.

54. Maraldi, "Recherche de la Parallaxe de Mars."

55. Fontenelle, *Entretiens*, 117.

56. Abetti, *The History of Astronomy*, 127.

57. G. D. Cassini, *Les Élémens de l'Astronomie*, 115–16.

58. Isaac Newton, *Philosophiae Naturalis Principia Mathematica*, 2:581, line 15 note.

59. Ibid.

60. Isaac Newton, *Opticks*, query 21 in all editions after 1717. See, e.g., the Dover reprint of the 1730 edition, p. 351.

61. Halley, "Some Remarks," 114. See also Bradley, *Miscellaneous Works and Correspondence*, pp. iv and 353.

62. Isaac Newton, *Philosophiae Naturalis Principia Mathematica*, 2:581.

63. Halley, *Phil. Trans.* 29, no. 348 (1716): 454–64.

64. Ibid., 455–56.

65. J. Cassini, "Recherche de la Parallaxe du Soleil"; Lalande, *Astronomie*, 2:414–15 (articles 1740–41). See also Maraldi, "De la Parallaxe de Mars," for Maraldi's effort of 1719.

66. Paris, Observatoire MSS. A6, 9, sheet no. 61, 20, 0.

67. Whiston, *Astronomical Lectures*, 85.

68. Many still learned the old values on their first exposure to this subject. See, e.g., Fontenelle, *Entretiens*, 117 n. 38. In Thomas Corneille's *Dictionnaire des Arts et des Sciences* of 1694, the Sun's size is given as 166 times the Earth's (2:409). The sizes of the other heavenly bodies are given accordingly.

69. *OCCH*, 21:801. I have taken the translation from the first English edition, *The Celestial Worlds Discovered* (1698), 138–39.

70. *OCCH*, 21:805; *The Celestial Worlds Discovered*, 139.

71. *OCCH*, 21:805, 807; *The Celestial Worlds Discovered*, 140–41.

72. *OCCH*, 21:809; *The Celestial Worlds Discovered*, 145.

73. Hooke, *An Attempt to Prove the Motion of the Earth*, 23–25. See also Williams, *Attempts to Measure Annual Stellar Parallax*, 21–27.

74. Registre, 9:111r.

75. Flamsteed to John Wallis, 20 December 1698, in Wallis, *Opera Mathematica*, 3:701–8.

76. J. Cassini, "Reflexions sur une Lettre," 177–83.

77. Whiston, *Astronomical Lectures*, 28–35.

78. *OCCH*, 21:814–17.

79. Isaac Newton, *A Treatise of the System of the World*. See idem, *Mathematical Principles of Natural Philosophy and His System of the World*, 596.

80. *OCCH*, 21:817.

81. Roberts, *Phil. Trans.* 18, no. 209 (1694): 103.

Bibliography

Abers, Ernest S., and C. F. Kennel. "Commentary: The Role of Error in Ancient Methods for Determining the Solar Distance." In *The Copernican Achievement*, ed. Robert S. Westman, 130–36. Berkeley: University of California Press, 1975.

Abetti, Giorgio. *The History of Astronomy.* Translated by Betty Burr Abetti. London: Sidgwick and Jackson, 1954.

Académie Royale des Sciences. *Histoire de l'Académie Royale des Sciences (1666–1699).* 2 vols. Paris, 1733.

______. *Mémoires de l'Académie Royale des Sciences depuis 1666 jusqu'à 1699.* 9 vols., numbered 3–11. Paris, 1729–32.

______. Archives. "Registre des procès-verbaux des séances."

Ambassades du Roy de Siam envoyé à l'Excellence du Prince Maurice arrivé à la Haye le 10. Septemb. 1608. The Hague, 1608. Reprinted in Stillman Drake, *The Unsung Journalist and the Origin of the Telescope.* Los Angeles: Zeitlin and Verbrugge, 1976.

Applebaum, Wilbur. "Horrocks." In *Dictionary of Scientific Biography.* 6:514–16. 16 vols. New York: Charles Scribner's Sons, 1970–80.

Applebaum, Wilbur, and Robert A. Hatch. "Boulliau, Mercator, and Horrocks's *Venus in Sole Visa*: Three Unpublished Letters." *Journal for the History of Astronomy* 14(1983):166–79.

Archimedes. *The Works of Archimedes.* Translated by T. L. Heath. Cambridge: Cambridge University Press, 1897–1912. Reprint. New York: Dover, 1956.

Aristarchus. *Aristarchus of Samos on the Sizes and Distances of the Sun and Moon.* Translated by T. L. Heath. In *Aristarchus of Samos the Ancient Copernicus*, 351–411. Oxford: Clarendon Press, 1913.

Armitage, Angus. *Edmond Halley.* London: Nelson, 1966.

Bacon, Roger. *The Opus Maius of Roger Bacon.* Translated by Robert Belle Burke. 2 vols. Philadelphia: University of Pennsylvania Press, 1928.

Baily, Francis. *An Account of the Revd. John Flamsteed.* London, 1835. Reprint. London: Dawson's, 1966.

Barrettus, Lucius [Albertus Curtius]. *Historia Coelestis.* Augsburg, 1666.

al-Battānī. *Al-Battānī sive Albatenii Opus Astronomicum.* 3 vols. Edited and translated by C. A. Nallino. *Pubblicazioni del Reale Osservatorio di Brera in Milano* 49, part 1 (1903); part 2 (1907); part 3 (1899).

Beaver, Donald DeB. "Bernard Walther: Innovator in Astronomical Observation." *Journal for the History of Astronomy* 1 (1970):39–43.

Benjamin, Francis S., Jr., and G. J. Toomer, eds. *Campanus of Novara and Medieval Planetary Theory.* Madison: University of Wisconsin Press, 1971.

Bennett, J. A. *The Mathematical Science of Christopher Wren.* Cambridge: Cambridge University Press, 1982.

Bentley, Richard. *Eight Lectures on Atheism.* London, 1692–93. Reprint. New York and London: Garland, 1976.

Bianchini, Francesco. "Nova Methodus Cassiniana, observandi Parallaxes & Distantias Planetarum a Terra, tentata Romae a Clarissimo Abbate Francisco Blanchino, transmissa Lipsiam ab Illustriss. Emanuele a Schelstraten, Bibliothecae Vaticanae Praefecto." *Acta Eruditorum*, October 1685, 470–78.

al-Bīrūnī. *The Determination of the Coordinates of Positions for the Correction of the Distances between Cities.* Translated by Jamil Ali. Beirut: American University of Beirut, 1967.

Boulliau, Ismael. *Astronomia Philolaica.* Paris, 1645.

Bradley, James. *Miscellaneous Works and Correspondence of James Bradley.* Edited by Stephen P. Rigaud. Oxford, 1832; New York: Johnson Reprint Corp., 1972.

Brahe, Tycho. *Tychonis Brahe Opera Omnia.* Edited by J. L. E. Dreyer. 15 vols. Copenhagen, 1913–28. Reprint. Amsterdam: Swets and Zeitlinger, 1972.

Buchdahl, Gerd. "Methodological Aspects of Kepler's Theory of Refraction." *Studies in the History and Philosophy of Science* 3 (1972):265–98.

Campanus of Novara. *Theorica Planetarum.* In *Campanus of Novara and Medieval Planetary Theory.* Edited by Francis S. Benjamin, Jr., and G. J. Toomer. Madison: University of Wisconsin Press, 1971.

Carlos, Edward Stafford, trans. *The Sidereal Messenger of Galileo Galilei and a Part of the Preface to Kepler's Dioptrics.* London, 1880. Reprint. London: Dawson's, 1960.

Carmody, Francis J. "Leopold of Austria 'Li Compilacions de le Science des Estoilles,' Books I–III. Edited from Ms. French 613 of the Bibliothèque Nationale, with Notes and Glossary." *University of California Publications in Modern Philology* 33, part 2 (1947):i–iv, 35–102.

______. "Li Livres dou Tresor de Brunetto Latini." *University of California Publications in Modern Philology* 22 (1948).

______, ed. *The Astronomical Works of Thabit b. Qurra.* Berkeley: University of California Press, 1960.

______, ed. *Al-Farghani Differentie Scientie Astrorum.* Translated by John of Seville. Berkeley, Calif.: Privately printed, 1943.

Caspar, Max. *Bibliographia Kepleriana* (1936). 2d ed. Edited by Martha List. Munich: C. H. Beck, 1968.

______. *Kepler.* Translated by C. Doris Hellman. New York: Abelard-Schuman, 1959.

Cassini, Giovanni Domenico. *De l'Origine et du Progrès de l'Astronomie, et de son Usage dans la Géographie et dans la Navigation* (1693). In *Mémoires de l'Académie Royale des Sciences depuis 1666 jusqu'à 1699.* 8:1–52. Paris, 1730.

______. *De Solaribus hypothesibus et refractionibus epistolae tres.* Bologna, 1666. Reprinted in *Miscellanea Italica Physico-Mathematica*, ed. Gaudenzio Roberti, 283–340. Bologna, 1692.

______. *Les Élémens de l'Astronomie, Verifiés par le rapport des Tables aux Observations de M. Richer, faites en l'Isle de Caïenne* (1684). In *Mémoires de l'Académie Royale des Sciences depuis 1666 jusqu'à 1699.* 8:53–117. Paris, 1730.

______. *La Meridiana del Tempio di S. Petronio. Tirata, e preparata per l'Osservazioni Astronomiche l'Anno 1655. Rivista, e restaurata l'Anno 1695.* Bologna, 1695.

______. *Observations Astronomiques faites en Divers Endroits du Royaume pendant l'Année 1672* (1693). In *Mémoires de l'Académie Royale des Sciences depuis 1666 jusquà 1699.* 7, part 1:349–74.

______. *Observations sur la Comète qui a paru au Mois de Décembre 1680. et en Janvier 1681.* Paris, 1681.

______. *Specimen Observationum Bononiensum.* Bologna, 1656.

Cassini, Jacques. "Recherche de la Parallaxe du Soleil par l'Observation de Mars, au temps de son Oppostion avec le Soleil, de l'Année 1736." In *Histoire de l'Académie Royale des Sciences avec les Mémoires de Mathématique et de Physique tirés des Registres de cette Académie. Mémoires*. 1739, 197–220.

———. "Réflexions sur une Lettre de M. Flamsteed à M. Wallis, touchant la Parallaxe annuelle de l'Étoile Polaire." In *Histoire de l'Académie Royale des Sciences avec les Mémoires de Mathématique et de Physique tirés des Registres de cette Académie. Mémoires*. 1699, 177–83.

Caxton, William. *Caxton's Mirrour of the World*. Edited by O. H. Prior. Early English Text Society, extra series no. 110. 1913.

Christmann, Jacob. *Nodus Gordius ex Doctrina Sinuum explicatus. Accedit Appendix Observationum, quae per Radium Artificiosum habitae sunt, circa Saturnum, Iovem, et Lucidiores Stellas Affixas*. Heidelberg, 1612.

Clavius, Christopher. *Christophori Clavii Opera Mathematica*. 6 vols. Mainz, 1611–12.

Cohen, I. Bernard. "Perfect Numbers in the Copernian System: Rheticus and Huygens." In *Science and History: Studies in Honor of Edward Rosen*, eds. Erna Hilfstein, Paweł Czartoryski, and Frank D. Grande, 419–25. *Studia Copernicana*, vol. 16. Warsaw: Ossolineum, Polish Academy of Sciences Press, 1978.

Colombe, Ludovico delle. *Contro il Moto della Terra* (1610). In *Le Opere di Galileo Galilei*, Edizione Nazionale, ed. Antonio Favara, vol. 3, part 1, 253–90. Florence, 1890–1909; reprinted 1929–39, 1964–66.

Copernicus, Nicholas. *De Revolutionibus Orbium Caelestium Libri Sex*. Vol. 2, *Nikolaus Kopernikus Gesamtausgabe*. 2 vols. Munich: Oldenbourg, 1944–49.

———. *Nicholas Copernicus on the Revolutions*. Translated by Edward Rosen. Vol. 2, *Nicholas Copernicus Complete Works*. Warsaw and Cracow: Polish Scientific Publishers; London: Macmillan, 1972–. Published separately, Baltimore: Johns Hopkins University Press, 1978.

Corneille, Thomas. *Dictionnaire des Arts et des Sciences*. 2 vols. Paris, 1694; Geneva: Slathine Reprints, 1968.

Cornford, Francis MacDonald. *Plato's Cosmology: The Timaeus of Plato*. New York: Harcourt, Brace, 1937.

Curtius, Albertus. *See* Barrettus, Lucius.

Cysat, Johannes. Letter to Johannes Lanz, 15 December 1631. Dillingen, Studienbibliothek. MSS. 2°, vol. 247, fols. 378–80.

d'Ailly, Pierre. *Ymago Mundi*. In *A Source Book in Medieval Science*, ed. Edward Grant, 630–39. Cambridge: Harvard University Press, 1974.

Dante Alighieri. *Dante's Convivio*. Translated by W. W. Jackson. Oxford: Clarendon Press, 1909.

———. *The Divine Comedy*. Translated by C. S. Singleton. 3 vols. in 6 parts. Princeton: Princeton University Press, 1970–75.

D'Evelyn, C., and A. J. Mill, eds. *The South English Legendary*. 3 vols. Early English Text Society, nos. 235, 236 (1956), no. 244 (1959).

Dicks, D. R. *Early Greek Astronomy to Aristotle*. Ithaca, N.Y.: Cornell University Press, 1970.

———. "Solstices, Equinoxes, and the Presocratics." *Journal of Hellenic Studies* 86 (1966):26–40.

Dictionary of Scientific Biography. 16 vols. New York: Charles Scribner's Sons, 1970–80.

Diogenes Laertius. *Lives of the Eminent Philosophers*. Translated by R. D. Dicks. London: Heinemann and New York: G. P. Putnam's Sons, 1925.

Drake, Stillman. *Galileo at Work: His Scientific Biography*. Chicago: University of Chicago Press, 1978.

______. "Galileo's First Telescopes at Padua and Venice." *Isis* 50 (1959): 245–54.

______. "Galileo's First Telescopic Observations." *Journal for the History of Astronomy* 7 (1976): 153–68.

______. *The Unsung Journalist and the Origin of the Telescope*. Los Angeles: Zeitlin and Verbrugge, 1976.

______, ed., trans. *Discoveries and Opinions of Galileo*. New York: Doubleday, 1957.

Drake, Stillman, and C. D. O'Malley, eds., trans. *The Controversy on the Comets of 1618*. Philadelphia: University of Pennsylvania Press, 1960.

Dreyer, J. L. E. *History of the Planetary Systems from Thales to Kepler*. Cambridge: Cambridge University Press, 1906. Reprinted as *A History of Astronomy from Thales to Kepler.* New York: Dover, 1953.

______. "Mediaeval Astronomy." In *Studies in the History and Method of Science*, 2 vols., ed. Charles Singer, 2:102–20. Oxford: Clarendon Press, 1917–21. Reprinted in Robert Palter, ed. *Toward Modern Science*. 2 vols. 1:235–56. New York: Noonday Press, 1961.

______. *Tycho Brahe*. Edinburgh, 1890. Reprint. New York: Dover, 1963.

Einhard. *Annales Laurissensis et Einhardi*. In *Monumenta Germaniae Historica Scriptorum*. 1:124–218. Hanover, 1826.

Euclid. *The Thirteen Books of Euclid's Elements*. Translated by T. L. Heath. 3 vols. Cambridge: Cambridge University Press, 1908; 2d ed., 1926. Reprint. New York: Dover, 1956.

al-Farghānī. *Al-Farghani Differentie Scientie Astrorum*. Translated by John of Seville. Edited by Francis J. Carmody. Berkeley, Calif.: Privately printed, 1943.

______. *Continentur in hoc Libro Rudimenta Astronomica Alfragani*. Nuremberg, 1537.

Flamsteed, John. "Flamsteed Papers." Archive, Royal Greenwich Observatory, Herstmonceux.

______. *The Gresham Lectures of John Flamsteed*. Edited by Eric G. Forbes. London: Mansell, 1975.

______. *Historia Coelestis Britannica*. 3 vols. London, 1725.

Fontana, Francesco. *Novae Coelestium Terrestriumque Rerum Observationes*. Naples, 1646.

Fontenelle, Bernard le Bovier de. *Entretiens sur la Pluralité des Mondes. Digression sur les Anciens et les Modernes*. Edited by Robert Shackleton. Oxford: Clarendon Press, 1955.

Forbes, Eric G., ed. *The Gresham Lectures of John Flamsteed*. London: Mansell, 1975.

Galilei, Galileo. *Dialogue concerning the Two Chief World Systems—Ptolemaic & Copernican*. Translated by Stillman Drake. Berkeley: University of California Press, 1967.

______. *Discoveries and Opinions of Galileo*. Translated by Stillman Drake. New York: Doubleday, 1957.

______. *Le Opere di Galileo Galilei*. Edizione Nazionale. 20 vols. Edited by Antonio Favaro. Florence, 1890–1909; reprinted 1929–39, 1964–66.

Gassendi, Pierre. *Petri Gassendi Opera Omnia*. 6 vols. Lyons, 1658.

Gaythorpe, Sidney B. "Horrocks's Observation of the Transit of Venus 1639 November 24 (O.S.)." *Journal of the British Astronomical Association* 47 (1936): 60–69; 64 (1954): 309–15.

Gingerich, Owen. "Dreyer and Tycho's World System." *Sky and Telescope* 64, no. 2 (August 1982): 138–40.

______. "The Origins of Kepler's Third Law." *Vistas in Astronomy* 18 (1975): 595–601.

Goldstein, Bernard R. *The Arabic Version of Ptolemy's "Planetary Hypotheses."* American Philosophical Society, *Transactions* 57, part 4 (1967).
______. *The Astronomical Tables of Levi ben Gerson*. Hamden, Conn.: Archon Books, 1974.
______. "Some Medieval Reports on Venus and Mercury Transits." *Centaurus* 14 (1969): 49–59.
______. "Theory and Observation in Medieval Astronomy." *Isis* 63 (1972): 39–47.
Goldstein, Bernard R., and Noel M. Swerdlow. "Planetary Distances and Sizes in an Anonymous Arabic Treatise Preserved in Bodleian MS. Marsh 621." *Centaurus* 15 (1970): 135–70.
Grafton, Anthony. "Michael Maestlin's Account of Copernican Planetary Theory." *Proceedings of the American Philosophical Society* 117 (1973): 523–50.
Grant, Edward. "Cosmology." In *Science in the Middle Ages*, ed. David C. Lindberg, 265–302. Chicago: University of Chicago Press, 1978.
______. *A Source Book in Medieval Science*. Cambridge: Harvard University Press, 1974.
Gregory, David. *Astronomiae Physicae & Geometricae Elementa*. Oxford, 1702.
______. *The Elements of Physical and Geometrical Astronomy*. 2 vols. London, 1726. New York: Johnson Reprint Corp., 1972.
Gregory, James. *Optica Promota*. London, 1663.
Grosseteste, Robert. *Die Philosophischen Werke des Robert Grosseteste*. Edited by L. Baur. Vol. 9, *Beitrage zur Geschichte der Philosophie des Mittelalters*. Münster i. W., 1912.
Gunther, R. T. *Early Science in Oxford*. 14 vols. Oxford: Privately printed, 1923–45.
Hall, A. R. "Newton on the Calculation of Central Forces." *Annals of Science* 13 (1957): 62–71.
Hall, A. R., and M. B. Hall. *The Correspondence of Henry Oldenburg*. Madison: University of Wisconsin Press and London: Mansell, 1963–.
Halley, Edmond. *Catalogus Stellarum Australium*. London, 1679.
______. "De Visibili Conjunctione Inferiorum Planetarum cum Sole, Dissertatio Astronomica." *Philosophical Transactions* 17, no. 193 (1691): 511–22.
______. "Methodus Singularis qua Solis Parallaxis sive Distantia a Terra, ope Veneris intra Solem conspicienda, tuto determinari poterit." *Philosophical Transactions* 29, no. 348 (1716): 454–64.
______. "Some Remarks upon the Method of Observing the Differences of Right Ascension and Declination by Cross Hairs in a Telescope." *Philosophical Transactions* 31, no. 366 (1720): 113–16.
Hartner, Willy. "Medieval Views on Cosmic Dimensions and Ptolemy's Kitāb al-Manshūrāt." In *Mélanges Alexandre Koyré*, 2 vols. 1:254–82. Paris: Hermann, 1964. Reprinted in *Oriens Occidens*, 319–48. Hildesheim: Georg Olms, 1968.
______. "Nasir al-Dīn al-Tūsī's Lunar Theory." *Physis* 11 (1969): 287–304.
______. "Ptolemy and Ibn Yūnus on Solar Parallax." *Archives Internationales d'Histoire des Sciences* 30 (1980): 5–26.
Hatch, Robert A. *See* Applebaum and Hatch.
Heath, T. L. *Aristarchus of Samos the Ancient Copernicus*. Oxford: Clarendon Press, 1913.
______, trans. *The Thirteen Books of Euclid's Elements*. 3 vols. Cambridge: Cambridge University Press, 1908; 2d ed., 1926. Reprint. New York: Dover, 1956.
______, trans. *The Works of Archimedes*. Cambridge: Cambridge University Press, 1897–1912. Reprint. New York: Dover, 1956–?

Hellman, C. Doris, and Noel M. Swerdlow. "Peurbach." *Dictionary of Scientific Biography*. 15:473–79.
Henderson, Janice A. "Erasmus Reinhold's Determination of the Distance of the Sun from the Earth." In *The Copernican Achievement*, ed. Robert S. Westman, 108–29. Berkeley: University of California Press, 1975.
______. *On the Distances between Sun, Moon, and Earth according to Ptolemy, Copernicus, and Reinhold*. Ph.D. diss., Yale University, 1973.
Hérigone, Pierre. *Cursus Mathematicus*. 2d ed. 6 vols. Paris, 1644.
Herivel, John. *The Background to Newton's Principia*. Oxford: Clarendon Press, 1965.
Herrmann, Dieter B. *Kosmische Weiten. Geschichte der Entfernungsmessung im Weltall*. Leipzig: Barth, 1977.
Hevelius, Johannes. *Mercurius in Sole Visus*. Gdansk, 1662.
______. *Selenographia*. Gdansk, 1647.
Hooke, Robert. *An Attempt to Prove the Motion of the Earth by Observations*. London, 1674. Reprinted in R. T. Gunther, *Early Science in Oxford*. 8:1–28. Oxford: Privately printed, 1923–45.
Horrocks, Jeremiah. *Jeremiae Horrocci Opera Posthuma*. Edited by John Wallis. London, 1673.
______. *The Transit of Venus across the Sun: a Translation of the Celebrated Discourse thereupon, by the Rev. Jeremiah Horrox . . . to which is prefixed a Memoir of his Life and Labours by the Rev. Arundell Blount Whatton*. London, 1859, 1868.
______. *Venus in Sole Visa*. In J. Hevelius, *Mercurius in Sole Visus*, 111–45. Gdansk, 1662.
Hortensius, Martinus. *Martini Hortensi Delfensis Dissertatio de Mercurio in Sole Viso et Venere Invisa*. Leyden, 1633.
Hoyer, Ulrich. "Über die Unvereinbarkeit der drei Keplerschen Gesetze mit der Aristotelischen Mechanik." *Centaurus* 20 (1976):196–209.
Humbert, Pierre. "A Propos du Passage de Mercure 1631." *Revue d'Histoire des Sciences et de leurs Applications* 3 (1950):27–31.
Huygens, Christiaan. *The Celestial Worlds Discovered*. London, 1698. Reprint. London: Frank Cass, 1968.
______. *Oeuvres Complètes de Christiaan Huygens*. 22 vols. The Hague: Martinus Nijhoff, 1888–1950.
Kahn, C. H. *Anaximander and the Origin of Greek Cosmology*. New York: Columbia University Press, 1960.
Kepler, Johannes. *Epitome of Copernican Astronomy: IV and V*. Translated by Charles Glenn Wallis. In *Great Books of the Western World*, vol. 16, *Ptolemy, Copernicus, Kepler*, 845–1004. Chicago, London, and Toronto: Encyclopaedia Britannica, Inc., 1952.
______. *Joannis Kepleri Astronomi Opera Omnia*. Edited by C. Frisch. 8 vols. Frankfurt: Erlangen, 1858–70.
______. *Johannes Kepler Gesammelte Werke*. Vols. 1–10, 13–19. Munich: C. H. Beck, 1937–.
______. *Kepler's Conversation with Galileo's Sidereal Messenger*. Translated by Edward Rosen. New York: Johnson Reprint Corp., 1965.
______. *Kepler's Somnium*. Translated by Edward Rosen. Madison: University of Wisconsin Press, 1967.
______. *Mysterium Cosmographicum: The Secret of the World*. Translated by Alistair M. Duncan. New York: Abaris Books, 1981.
Koyré, Alexandre. *The Astronomical Revolution*. Translated by R. E. W. Madison. Ithaca, N.Y.: Cornell University Press, 1973.

———. *From the Closed World to the Infinite Universe*. Baltimore: Johns Hopkins Press, 1957; New York: Harper & Row, 1958.

Kremer, Richard L. "The Use of Walther's Astronomical Observations: Theory and Observation in Early Modern Astronomy." *Journal for the History of Astronomy* 12 (1981): 124–32.

Kren, Claudia. "Homocentric Astronomy in the Latin West: The *De Reprobatione Eccentricorum et Epiciclorum* of Henry of Hesse." *Isis* 59 (1968): 269–81.

Kuhn, Thomas S. *The Copernican Revolution*. Cambridge: Harvard University Press, 1957.

Lalande, Joseph-Jerôme le Français de. *Astronomie*. 2d ed. 4 vols. Paris, 1771–81.

Lansberge, Philip van. *Commentationes in Motum Terrae Diurnum, et Annuum*. Translated by M. Hortensius. Middelburg, 1630.

———. *Tabulae Motuum Coelestium Perpetuae*. Middelburg, 1632.

Latini, Brunetto. *Li Livres dou Tresor*. In Francis J. Carmody, ed., "Li Livres dou Tresor de Brunetto Latini." *University of California Publications in Modern Philology* 22 (1948).

Le Monnier, Pierre-Charles. *Histoire Céleste*. Paris, 1741.

Lewis, C. S. *The Discarded Image. An Introduction to Medieval and Renaissance Literature*. Cambridge: Cambridge University Press, 1964.

Lindberg, David C., ed. *Science in the Middle Ages*. Chicago: University of Chicago Press, 1978.

Locher, Johann Georg. *Disquisitiones Mathematicae de Controversiis et Novitatibus Astronomicis. Quas sub Praesidio Christophori Scheineri . . . publice disputandas posuit propugnavit . . . Johannes Georgius Locher*. Ingolstadt, 1614.

Lovejoy, Arthur O. *The Great Chain of Being*. Cambridge: Harvard University Press, 1936, 1948.

Maeyama, Y. "The Historical Development of Solar Theories in the Late Sixteenth and Seventeenth Centuries." *Vistas in Astronomy* 16 (1974): 35–60.

Magini, Giovanni Antonio. *Novae Coelestium Orbium Theoricae Congruentes cum Observationibus N. Copernici*. Venice, 1589.

Maimonides, Moses. *The Guide of the Perplexed*. Translated by S. Pines. Chicago: University of Chicago Press, 1963.

Malvasia, Cornelius. *Ephemerides Novissimae Motuum Coelestium*. Modena, 1662.

Maraldi, Giacomo. "De la Parallaxe de Mars." In *Histoire de l'Académie Royale des Sciences avec les Mémoires de Mathématique et de Physique tirés des Registres de cette Académie. Mémoires*, 1722, 216–29.

———. "Recherche de la Parallaxe de Mars." In *Histoire de l'Académie Royale des Sciences avec les Mémoires de Mathématique et de Physique tirés des Registres de cette Académie. Mémoires*, 1706, 69–74.

Marius, Simon. *Mundus Iovialis*. Translated by A. O. Prickard. In "The 'Mundus Jovialis' of Simon Marius." *The Observatory* 39 (1916): 367–81, 403–12, 443–52, 498–503.

McKeon, Robert M. "Les Débuts de l'Astronomie de Précision. I. Histoire de la Réalisation du Micromètre Astronomique." *Physis* 13 (1971): 225–88.

Mill, A. J., and C. D'Evelyn, eds. *The South English Legendary*. 3 vols. Early English Text Society, nos. 235, 236 (1956), no. 244 (1959).

Mims, Stewart L. *Colbert's West India Policy*. New Haven: Yale University Press, 1912.

Neugebauer, Otto. *The Exact Sciences in Antiquity*. 2d ed. Providence, R.I.: Brown University Press, 1957.

______. *A History of Ancient Mathematical Astronomy*. 3 parts. New York, Heidelberg, and Berlin: Springer-Verlag, 1975.
Newton, Isaac. *The Correspondence of Isaac Newton*. Edited by H. W. Turnbull, J. F. Scott, A. R. Hall, and Laura Tilling. 7 vols. Cambridge: Cambridge University Press, 1959–77.
______. *Isaac Newton's Philosophiae Naturalis Principia Mathematica*. Edited by Alexandre Koyré, I. Bernard Cohen, and Anne Whitman. 2 vols. Cambridge, Mass.: Harvard University Press; Cambridge: Cambridge University Press, 1972.
______. *Opticks*. 4th ed. London, 1730. Reprint. New York: Dover, 1952.
______. *Sir Isaac Newton's Mathematical Principles of Natural Philosophy and His System of the World*. Edited by Florian Cajori. Berkeley: University of California Press, 1934, 1946.
______. *A Treatise of the System of the World*. London, 1728.
Newton, Robert R. *The Crime of Claudius Ptolemy*. Baltimore: Johns Hopkins University Press, 1977.
Nicolson, Marjorie. *Science and Imagination*. Ithaca, N.Y.: Cornell University Press, 1956.
North, John D. "Thomas Harriot and the First Telescopic Observations of Sunspots." In *Thomas Harriot: Renaissance Scientist*, ed. John W. Shirley, 129–65. Oxford: Clarendon Press, 1974.
Olmsted, John. "The Scientific Expedition of Jean Richer to Cayenne." *Isis* 34 (1942): 117–28.
O'Malley, C. D., and Stillman Drake, eds., trans. *The Controversy on the Comets of 1618*. Philadelphia: University of Pennsylvania Press, 1960.
Orosius. *Historiarum adversus Paganos Libri VII*. Edited by C. Zangmeister. Vienna, 1882; New York: Johnson Reprint Corp., 1966.
Orr, M. A. *Dante and the Early Astronomers*. London, 1913. Reprint. London: Allan Wingate, 1966.
Osborne, Catherine. "Archimedes on the Dimensions of the Cosmos." *Isis* 74 (1983): 234–42.
Pannekoek, Antonie. *A History of Astronomy*. London: Allen & Unwin; New York: Barnes and Noble, 1961.
Pedersen, Olaf. "Astronomy." In *Science in the Middle Ages*, ed. David C. Lindberg, 303–37. Chicago: University of Chicago Press, 1978.
______. *A Survey of the Almagest*. Odense: Odense University Press, 1974.
Peurbach, Johannes. *Theoricae Novae Planetarum*. Nuremberg, 1474.
Picard, Jean. *Mesure de la Terre* (1671). In *Mêmoires de l' Académie Royale des Sciences depuis 1666 jusqu'à 1699*. 7, part 1: 133–90. Paris, 1729.
______. *Voyage d'Uranibourg, ou Observations Astronomiques faites au Danemarck* (1680). In *Mêmoires de la'Académie Royale des Sciences depuis 1666 jusqu'à 1699*. 7, part 1: 191–230. Paris, 1729.
Pingré, Alexandre-Guy. *Annales Célestes du Dix-Septième Siècle*. Edited by Guillaume Bigourdan. Paris: Gauthiers-Villars, 1901.
Plato. *Plato's Cosmology: The Timaeus of Plato*. Translated by Francis MacDonald Cornford. New York: Harcourt, Brace, 1937.
Proclus. *In Platonis Timaeus Commentaria*. Edited by E. Diehls. 3 vols. Leipzig: Teubner, 1903–6. Reprint. Amsterdam: Hakkert, 1965.
______. *Procli Diadochi Hypotyposis Astronomicarum Positionum*. Edited and translated by C. Manitius. Leipzig: Teubner, 1909. Reprint. Stuttgart: Teubner, 1974.
Ptolemy. *Claudii Ptolemaei Opera quae Exstant Omnia*. Edited by J. L. Heiberg, F. Boll, and Æ. Boer. 3 vols. Leipzig: Teubner, 1893–1952.

———. *Handbuch der Astronomie*. Translated by K. Manitius. 2 vols. Leipzig: Teubner, 1912–13, 1963.
———. *Planetary Hypotheses*. *See* Goldstein, Bernard R., 1967.
Regiomontanus [Johannes Müller]. *Epytoma Joannis de monte regio in almagestum ptolemei*. Venice, 1496.
Reinhold, Erasmus. *Prutenicae tabulae coelestium motuum*. Tubingen, 1551.
Rheita, Antonius Maria Schyrlaeus de. *Oculus Enoch et Eliae, sive Radius Sidereomysticus*. Antwerp, 1645.
Ricciolo, Giovanni Battista. *Almagestum Novum*. 2 parts, Bologna, 1651.
Richer, Jean. *Observations Astronomiques et Physiques faites en l'Isle de Caienne* (1679). In *Mémoires de l'Académie Royale des Sciences depuis 1666 jusqu'à 1699*. 7, part 1: 231–326. Paris, 1729.
Rigaud, Stephen P., ed. *Correspondence of Scientific Men*. 2 vols. Oxford, 1841. Reprint. Hildesheim: Georg Olms, 1965.
Risner, Friedrich, ed. *Opticae Thesaurus*. Basle, 1572; New York: Johnson Reprint Corp., 1972.
Roberts, Francis. "Concerning the Distance of the Fixed Stars." *Philosophical Transactions* 18, no. 209 (1694): 101–3.
Roemer, Ole. "Démonstration touchant le Mouvement de la Lumière." *Journal des Sçavans*, 7 December 1676, 233–36.
———. "A Demonstration concerning the Motion of Light, communicated from Paris, in the *Journal des Sçavans*, and here made English." *Philosophical Transactions* 12, no. 136 (1677): 893–94.
Rosen, Edward. "Copernicus' Spheres and Epicycles." *Archives Internationales d'Histoire des Sciences* 25 (1975): 82–92.
———, trans. *Kepler's Conversation with Galileo's Sidereal Messenger*. New York: Johnson Reprint Corp., 1965.
———, trans. *Kepler's Somnium*. Madison: University of Wisconsin Press, 1967.
———, trans. *Nicholas Copernicus on the Revolutions*. Vol 2, *Nicholas Copernicus Complete Works*. Warsaw and Cracow: Polish Scientific Publishers; London: Macmillan, 1972–. Published separately, Baltimore: Johns Hopkins University Press, 1978.
———. "Regiomontanus." *Dictionary of Scientific Biography*, 11:348–52.
———. "Reply to N. Swerdlow." *Archives Internationales d'Histoire des Sciences* 26 (1976): 302–4.
———. "Richer." *Dictionary of Scientific Biography*, 11:423–25.
———. "When Did Galileo Make His First Telescope?" *Centaurus* 2 (1951): 44–51.
Sacrobosco, Johannes. *See* Thorndike, Lynn.
Saliba, George. "The First Non-Ptolemaic Astronomy at the Maraghah School." *Isis* 70 (1979): 571–76.
Sarton, George. "Early Observations of the Sunspots?" *Isis* 37 (1947): 69–71.
Scheiner, Christoph. *Apelles Latens post Tabulam de Maculis Solaribus et Stellis circa Iovem Errantibus Accuratior Disquisitio ad Marcum Velserum* (1612). In *Le Opere di Galileo Galilei*, Edizione Nazionale, ed. Antonio Favaro, 5:37–70. Florence, 1890–1909; reprinted 1929–39, 1964–66.
———. *Disquisitiones Mathematicae de Controversiis et Novitatibus Astronomicis. Quas sub Praesidio Christophori Scheineri . . . publice disputandas posuit propugnavit . . . Johannes Georgius Locher*. Ingolstadt, 1614.
———. *Tres Epistolae de Maculis Solaribus scriptae ad Marcum Velserum* (1612). In *Le Opere di Galileo Galilei*, Edizione Nazionale, ed. Antonio Favaro, 5:23–33. Florence, 1890–1909; reprinted 1929–39, 1964–66.
Schickard, Wilhelm. *W. Schickardi pars Responsi ad Epistolas P. Gassendi . . . de Mercurio sub Sole Viso*. Tübingen, 1632.

Shapiro, Alan E. "Archimedes's Measurement of the Sun's Apparent Diameter." *Journal for the History of Astronomy* 6 (1975): 75–83.

Sherburne, Edward. *The Sphere of Marcus Manilius made an English Poem: with Annotations and an Astronomical Appendix*. London, 1675.

Shirley, John W., ed. *Thomas Harriot: Renaissance Scientist*. Oxford: Clarendon Press, 1974.

______. "Thomas Harriot's Lunar Observations." In *Science and History: Studies in Honor of Edward Rosen*, eds. Erna Hilfstein, Paweł Czartoryski, and Frank D. Grande, 283–308. *Studia Copernicana*, vol. 16. Warsaw: Ossolineum, Polish Academy of Sciences Press, 1978.

Simplicius. *In Aristotelis de Caelo Commentaria*. Edited by J. L. Heiberg. Vol. 7, *Commentaria in Aristotelem Graeca*. 23 vols. Berlin, 1882–1909.

Streete, Thomas. *Astronomia Carolina*. London, 1661.

______. *An Appendix to Astronomia Carolina*. London, 1664.

______. *Examen Examinatum: or Wing's Examination of Astronomia Carolina Examined*. London, 1667.

Swerdlow, Noel M. "Al-Battānī's Determination of Solar Distance." *Centaurus* 17 (1972): 97–105.

______. "Hipparchus on the Distance of the Sun." *Centaurus* 14 (1969): 287–305.

______. "Pseudodoxia Copernicana: or Enquiries into very many received Tenets and commonly presumed Truths, mostly concerning Spheres." *Archives Internationales d'Histoire des Sciences* 26 (1976): 108–58.

______. *Ptolemy's Theory of the Distances and Sizes of the Planets: A Study in the Scientific Foundation of Medieval Cosmology*. Ph.D. diss., Yale University, 1968.

Swerdlow, Noel M., and Bernard R. Goldstein. "Planetary Distances and Sizes in an Anonymous Arabic Treatise Preserved in Bodleian MS. Marsh 621." *Centaurus* 15 (1970): 135–70.

Swerdlow, Noel M., and C. Doris Hellman. "Peurbach." *Dictionary of Scientific Biography*, 15:473–79.

Tannery, Paul. *Recherches sur l'Histoire de l'Astronomie Ancienne*. Paris, 1893. Reprint. New York: Arno Press, 1976.

Taton, René. "Cassini, Gian Domenico." *Dictionary of Scientific Biography*, 3: 100–104.

Thābit ibn Qurra. *De Hiis que Indigent Expositione antequam legatur Almagesti*. Translated by Gerard of Cremona. In *The Astronomical Works of Thabit b. Qurra*, ed. Francis J. Carmody, 131–39. Berkeley: University of California Press, 1960.

______. *De Quantitatibus Stellarum et Planetarum et Proportio Terre*. Translated by Gerard of Cremona. In *The Astronomical Works of Thabit b. Qurra*, ed. Francis J. Carmody, 145–48. Berkeley: University of California Press, 1960.

Thorndike, Lynn, ed., trans. *The Sphere of Sacrobosco and Its Commentators*. Chicago: University of Chicago Press, 1949.

Toomer, G. J. "Hipparchus." *Dictionary of Scientific Biography*, 15:207–24.

______. "Hipparchus on the Distances of the Sun and Moon." *Archive for History of Exact Sciences* 14 (1975): 126–42.

Toomer, G. J., and Francis S. Benjamin, Jr., eds., trans. *Campanus of Novara and Medieval Planetary Theory*. Madison: University of Wisconsin Press, 1971.

Toynbee, Paget. "Dante's Obligation to Alfraganus in the *Vita Nuova* and *Convivio*," *Romania* 24 (1895): 413–32.

Van Helden, Albert. " 'Annulo Cingitur': The Solution of the Problem of Saturn." *Journal for the History of Astronomy* 5 (1974): 155–74.

______. "The Importance of the Transit of Mercury of 1631." *Journal for the History of Astronomy* 7 (1976): 1–10.

———. *The Invention of the Telescope*. American Philosophical Society, *Transactions* 67, part 4 (1977).

———. "Roemer's Speed of Light." *Journal for the History of Astronomy* 14 (1983).

———. "The Telescope in the Seventeenth Century." *Isis* 65 (1974): 38–58.

Wallace, William. *Galileo's Early Notebooks: The Physical Questions*. Notre Dame: Notre Dame University Press, 1977.

Wallis, John. *Opera Mathematica*. 3 vols. Oxford, 1695–99.

Wendelin, Gottfried. *Eclipses Lunares ab Anno M.D. LXXIII. ad M.DC.XLIII. Observatae quibus Tabulae Atlanticae superstruuntur earumque Idea proponitur*. Antwerp, 1644.

———. *Loxias seu de Obliquitate Solis Diatriba*. Antwerp, 1626.

Westman, Robert S., ed. *The Copernican Achievement*. Berkeley: University of California Press, 1975.

Whiston, William. *Astronomical Lectures*. London, 1728; New York: Johnson Reprint Corp., 1972.

———. *A New Theory of the Earth*. London, 1696, 1737.

———. *Praelectiones Astronomicae*. Cambridge, 1707.

Williams, M. E. W. *Attempts to Measure Annual Stellar Parallax: Hooke to Bessel*. Ph.D. diss., Imperial College, University of London, 1981.

———. "Flamsteed's Alleged Measurement of Annual Parallax for the Pole Star." *Journal for the History of Astronomy* 10 (1979): 102–16.

Wilson, Curtis A. "Horrocks, Harmonies and the Exactitude of Kepler's Third Law." In *Science and History: Studies in Honor of Edward Rosen*, ed. Erna Hilfstein, Paweł Czartoryski, and Frank D. Grande, 235–59. *Studia Copernicana*, vol. 16. Warsaw: Ossolineum, Polish Academy of Sciences Press, 1978.

Wing, Vincent. *Astronomia Britannica*. London, 1669.

———. *Examen Astronomiae Carolinae: T. S.* London, 1665.

———. *'Ολύμπια Δώματα or, an Almanack and Prognostication for the Year of our Lord 1663*. London, 1663.

Zinner, Ernst. *Entstehung und Ausbreitung der Coppernicanischen Lehre*. Sitzungsbericht der Physikalisch-medizinischen Sozietät zu Erlangen, vol. 74. Erlangen: Max Mencke, 1943. Reprint. Vaduz, Liechtenstein: Topos, 1978.

Index

POINT LOMA NAZARENE COLLEGE
RYAN LIBRARY